Who's Afraid of Big Blue?

How Companies Are Challenging IBM— and Winning

 Regis McKenna

Addison-Wesley Publishing Company, Inc.
Reading, Massachusetts • Menlo Park, California • New York
Don Mills, Ontario • Wokinghham, England • Amsterdam • Bonn
Sydney • Singapore • Tokyo • Madrid • San Juan

Library of Congress Cataloging-in-Publication Data

McKenna, Regis
    Who's afraid of Big Blue?

    Includes index.
     1.   International Business Machines Corporation.
2.   Computer industry—United States.   3.   Competition—
United States.   I. Title.
HD9696.C64I4852   1989   338.7'61004'0973   88–7493
ISBN 0–201–15574-5

Cover design by Mike Fender
Text design by Judy Ashkenaz/Total Concept Associates
Set in 11-point Times Roman by Total Concept Associates

ABCDEFGHIJ-DO-898
First printing, November 1988

Acknowledgments

The computer industry is probably the most complex and volatile of all industries. Everything in the industry is constantly changing. The people, the products, the technologies, the alliances, even the standards don't sit still for very long. I did not set out to "understand" the industry but, rather, to gain some insights into what makes successful computer companies tick. I had a lot of help.

During the last three years, many people have helped me gain insight into IBM and its challengers. I visited executives at many challenger companies, and all were very cooperative and generous with their time. Among those who shared their time and ideas were Gene Amdahl, Ken Olsen, John Cullinane, Ed de Castro, Rod Canion, Jim Treybig, John Sculley, Bob Miller, Steve Jobs, Joe Zemke, Dave Martin, Floyd Kvamme, Mike Swavely, Bill Campbell, Chuck Boesenberg, Max Hopper, Richard Dalton, and Jonathan Seybold. We relied upon the research and resources of Hambrecht & Quist, International Data Corporation, Dataquest, and the Gartner Group.

Mary Ann Gilpatrick worked with me for two years, conducting background research, interviewing people, and drafting raw material. As the manuscript came to completion, Mary Ann gave birth to her first child. Mitch Resnick, who edited my first book, *The Regis Touch,* turned a mountain of material into organized prose. Both Mary Ann and Mitch made invaluable contributions to this book.

Floyd Kvamme, my partner at Kleiner Perkins Caufield & Byers, reviewed early drafts of the book. As former head of marketing at Apple Computer and former president of National Advanced Systems, Floyd contributed important background and insights. John Doerr and Jim Lally, also partners at KPC&B, taught me a great deal about the computer business and gave me the idea for this book. Tom Moser, head of Peat Marwick's high-technology practice, used his networking skills to help me access many of the people we interviewed. Bob Henkel, a superb technology journalist and a good friend for many years, provided encouragement and did his usual tough job of criticism on the first manuscript.

A number of people contributed help at Regis McKenna Inc., including Lee James, Susan Parrish, Gayle Holste, Elizabeth Chaney, Glenn Helton, Rob Brownstein, Jane Anderson, and Estee Soloman. Pamela Kirkbride and her research staff provided fact checking.

Special thanks to my wife, Dianne. We sat side by side at our Macintoshes many late evenings, weekends, and on our vacations, each working away on a product of this ever-challenging revolution.

Finally, I would like to thank my publisher, Ann Dilworth of Addison-Wesley. Ann is like a good venture capitalist. She doesn't just make the deal and wait for the return. She worked this deal from the start, providing guidance, criticism, changes to the approach— and all the way, she kept encouraging me.

Contents

> *I saw great businesses become but a ghost of a name because someone thought they could be managed just as they were always managed, and though the management may have been most excellent in its day, its excellence consisted in its alertness to its day, and not its slavish following of its yesterdays.*
>
> Henry Ford, *My Life and Work*

Introduction

■■ IN THIS COUNTRY, we have a problem with bigness. We have never found a very good way to manage large organizations. As our federal government expanded in the 1960s and 1970s, it became mired in bureaucracy. And as our leading corporations grew into industrial giants, they often became inflexible and unresponsive.

This inability to deal with bigness is becoming a more serious problem as the pace of change accelerates. In the business world, things are changing faster than ever before. New technologies, new products, new companies, and new markets are created every day. Yet big organizations have not found a good way to cope with change: they are always slow to respond, and often unable to adjust.

This book is about ways of coping with change, particularly change in the competitive world of computing. It's about perception and reality, and the way new images of leadership are emerging in the computer industry. It's about competition. It is about the competition that is challenging IBM for leadership in the computer industry—leadership based not on size, but on innovation and the development of new markets.

During the past twenty years or so, IBM has built an awe-inspiring image. Many customers view IBM as a technological leader, an unbeatable competitor, an innovative supercompany. No matter that these perceptions are largely inaccurate: customers believe them. People who make their living observing, analyzing, and writing about the computer industry reinforce this image. When IBM makes an announcement, industry "experts" discount the chances of every competitor in IBM's shadow. They assume that IBM will not only succeed in the market, but will dominate it. Big Blue, as IBM is sometimes called, based on its size and the color of its computers, has taken on an air of invincibility. Many customers are scared to buy from other computer vendors, and many vendors are scared to challenge IBM directly.

I remember discussing the "IBM dilemma" over lunch with a friend a few years ago. My friend, a venture capitalist, asked me: "What are we going to do about IBM? They're stifling innovation, killing new investments, frightening off would-be entrepreneurs and investors!"

I left the restaurant considering what my friend had said. I decided he was wrong. He had voiced the common perception about IBM, not the reality. The truth is, there are more innovations, more investments, and more market opportunities in the computer industry than ever before. Unfortunately, industry observers, as well as potential customers, see the world of computers through the aura enveloping IBM. It's a powerful image, but it doesn't reflect what's really happening.

IBM probably isn't going away. It won't cease to be the elephant charging about disrupting everyone's life in the jungle. But life in the jungle is different today. There are many more predators, and the territory is partitioned into many smaller ecosystems, leaving less room for the elephant to roam. The elephant isn't an endangered species. But despite the dire forecasts of some industry soothsayers, the many other species roaming the jungle aren't endangered, either.

When we look underneath IBM's carefully maintained image, reality tells us that it's the *other* players in the computing world that really drive the industry these days. They do it either through innovation or through unique ways of doing business. It is IBM that reacts. Thus, to truly understand the world of computing, we must enter and understand the world of the challengers, the innovators: the people who really effect change in today's computer industry.

These challengers deal with the myths and the reality of IBM on a real-time competitive battlefield. They have respect for Big Blue, but they do not fear it. They have, in fact, come to understand not only how to compete, but how to compete successfully and win.

The challenger companies are tiny compared to IBM. In 1986, IBM made Apple's revenues in 13 days, and it achieved Apple's profits in 75 days. But it is these challengers that are changing the nature and the future of computing. The important changes in the industry are coming from Digital, Cray, Apple, Compaq, Tandem, Convex, Sun, Lotus, Microsoft, Hewlett-Packard, and many others— *not* IBM. These other companies drive IBM. They make it react. They force the slow-moving giant to make reluctant changes. Viewed this way, IBM doesn't look so invincible.

The world of computing is changing both rapidly and dramatically. The clear lens of reality shows us that IBM can no longer be everything to everybody; things have become too complex. In 1970, there were 340 computer companies. By 1987, there were more than 10,000 companies in the computer business. The exact number of companies in the computer business is difficult to determine; with the advent of the microprocessor, the computers-on-a-chip, computers have become truly ubiquitous. There are computers everywhere: in televisions, in cars, in microwave ovens.

The microprocessor can turn virtually anything into a computer. Each new generation of chips leads to electronic equipment that is "smarter" than before. Today, a convenience store clerk operating a "state-of-the-art checkout terminal" commands at least two to three times more computer memory than did a data-processing manager at a medium-sized bank in 1975. Not only has the cash register become a computer, so has the gas pump, the machine tool, the traffic light, the telephone, the bank teller, the typewriter, and so on. General Motors is in the computer business. RCA, GE, Ford, Westinghouse, Tektronix, TRW—all these companies are in the computer business in one way or another.

As technology has advanced, it has dramatically simplified computer design. The costs of providing computer solutions to business problems have plummeted. With less capital required to enter the market, small companies are rushing into the computer business, creating new niches, offering new technologies, and providing eager customers with cutting-edge products. Innovation in computing is occurring at a breakneck speed, creating new products, applications, and markets so fast that it is very difficult for a large company like IBM to plug all the leaks in the huge, ever-fragmenting marketplace.

Small companies are developing thousands of new programs to apply computers to specific tasks. More lines of computer software were written in the last five years than in all previous history. Today, with new microprocessor technology, computers can be tailored to specific tasks or specific industries without added cost to users. When a computer is tailored to a specific task, its performance can also be enhanced for that task. Now doctors, lawyers, teachers, engineers, chemists, scientists, architects, writers, journalists, pharmacists, accountants, churches, charities, banks, travel agencies, brokers, telemarketing agencies, and passport offices can "have it their way" with computers tailored to meet their particular needs.

Not so very long ago, only a few specially trained people knew the secret of unlocking the power of the computer. Today, children learn to use and program computers as quickly as they learn to ride bicycles. And many more people have access to computers. When large mainframe computers dominated the computer business, only large institutions could afford to buy and maintain computers. Control of the entire installed base of computers was in the hands of very few people. Today, there are millions of computers in homes, schools, and small businesses. And there are great opportunities for further growth. There are 3,300 colleges and universities, 87 million homes, and an estimated 25 million business in the United States alone—and more than half a million new businesses are started each year.

Every day new computer markets arc created—markets that begin to stake claims within what was once IBM's exclusive territory. When IBM got its start in computers back in the 1950s, the computer world consisted of large mainframe computers and nothing else. The mainframe was an awesome creation kept in the back room to do number crunching—very monolithic, very identifiable. During the last twenty years, however, the role of computers has been completely redefined. Beginning with Digital, Control Data, and Cray, companies recognized opportunities to solve problems not addressed by IBM. They expanded the marketplace by providing new solutions.

As IBM continued to cling to the familiar mainframe market, startups created entirely new markets for computing. Digital created the minicomputer market; Tandem created the market for nonstop computing and transaction processing; Apple, personal computers; Cray, the supercomputer; Apollo and Sun, workstations; Mentor and Daisy, design workstations for engineers; Convex, the minisupercomputer. I could name many more companies that created markets through innovation. Today, hundreds of new and old computer companies are creating new products, new markets, new opportunities.

IBM, having risen to power in the days of the mainframe, is facing a difficult adjustment. IBM has always looked at the computer business as a market-share battle instead of a market-creation opportunity. IBM was extraordinarily successful at grabbing the biggest slice of the mainframe pie. But creation of new markets is a very different game.

In general, IBM continues to see the world through mainframe eyes, focusing its strategies on protecting its own high-margin monoliths, rather than creating new products and new markets. Because of its size and its myopic view of the computer world, IBM was very slow to see that new technology was fragmenting its markets. As a result, IBM missed the opportunity to adapt its technology to the new markets, and to develop new distribution channels that could move the technology to untapped markets.

The industry watches with great interest as IBM struggles to regain its balance. IBM's role in the computer business is slowly eroding as smaller, less expensive, specialized machines replace Big Blue's general-purpose giants. The old formulas for dominating the computer industry no longer apply. *Bigger isn't always better.* In fact, big machines and big companies are at a distinct disadvantage in the new, fragmenting computer market.

As the computer industry expands and becomes more diversified, IBM is finding that it cannot be all things to all people. For one thing, it can't move quickly enough—not because it doesn't want to, but because it is anchored to its past machines and firmly entrenched habits. Its actions tend to be defensive and concentrated on protecting established turf, rather than breaking new territory. Decisions must work their way through an entangled bureaucracy, and they must fit into a complex product strategy. Using new technology to build a new generation of computers poses major problems. How much of the old should be incorporated into the new? Users have made huge investments in the old software and hardware. Although customers want faster, more powerful machines, they also want to protect their past investments. Such considerations complicate and prolong Big Blue's decision-making process.

New computer companies with advanced architectures are unburdened by old baggage. Without ties to obsolete installed bases, they can quickly and efficiently offer computers designed to meet current customer needs. As a result, smaller companies are splintering IBM's mainframe market. Some companies are selling specialized machines tailored to the needs of particular users, offering more cost-effective solutions than the general-purpose mainframe. Other challenger companies are producing souped-up scientific computers, skimming high-end users off the top of IBM's customer base. Still other companies are producing smaller computers that, when linked into networks, can do the job of a mainframe at a fraction of the cost. There are more than 150 challenger companies selling personal computers today. They, not IBM, are bringing new ideas and new technologies to that fast-growing market.

Our images of IBM and the computer market lag behind the times. IBM's computer dominance was established in an age when the real battle was resource against resource, monolith against monolith, slugging it out in a less competitive and slower moving technological world. Our enduring image of IBM was formed during this age of monoliths. Too often, people think of IBM as it was in those days, when its major competitors were the so-called BUNCH companies (Burroughs, Univac, NCR, Control Data, and Honeywell), along with RCA and General Electric. Big Blue eventually drove GE and RCA out of the computer business. IBM's strategic, monetary, and market might left a lasting impression. It established IBM's position. People cling to the image of IBM as it was in the 1960s and 1970s. But that image doesn't accurately reflect the real IBM of today.

The world has changed but many of our perceptions have not. Today, IBM is competing not with old-time monoliths, but with a multitude of imaginative, small, hungry, quick-acting competitors. Today's IBM is a much different company competing in a much different game—a game whose rules are no longer skewed in its favor.

❑❑ The Challengers

When you have a small company, a lot of fear, and great opportunity, you can do amazing things.

Thomas Watson Jr., long-time president of IBM, wrote these words. Watson was referring to the way in which IBM overcame Univac in the early days of the computer industry. But the words could apply equally well to the situation in the computer industry today. The only difference is that the tables are now turned: it is other companies that are doing the "amazing things," this time at IBM's expense.

During the past twenty-five years, I've worked with some of the most successful IBM challengers, giving them advice, helping them plan marketing strategies for competing with IBM. I have found that the computer world looks very different through the eyes of these challengers. They see a world full of opportunities. Rather than seeing old markets dominated by IBM, they see new markets dominated by no one—indeed, they see markets and technologies that others have yet to imagine.

I decided to write this book to share the challengers' vision of the computer industry. I wanted to explain how certain challengers are reinventing the industry and, in the process, breaking IBM's stranglehold on the computer market. And I wanted to help other companies learn how they, too, can compete against corporate giants. I already knew most of the challengers quite well. But I wanted to probe more deeply, to get a better understanding of their strategies and ideas. So I set out to have a series of informal talks with some of the most successful challengers.

In all, I interviewed several dozen industry leaders, people who, in my opinion, are representative of the challengers now driving the computer industry. I talked with computer industry veterans like Digital's Ken Olsen and Tandem's Jimmy Treybig, and with relative newcomers like Sun's Scott McNealy. I talked with leaders of the personal computer industry like Apple's John Sculley and Compaq's Rod Canion. I talked with software industry leaders like Cullinet's John Cullinane. And I talked with IBM alumni like Gene Amdahl, Joe Zemke, and Bob Miller. Among the others: Hewlett-Packard president John Young, Honeywell Bull chairman Jacques Stern, Convex founders Bob Paluck and Steve Wallach, Digital Equipment pioneer Gordon Bell. Many of the people I talked to were among the founders of their companies. Olsen, Treybig, McNealy, Canion, Paluck, Wallach, and Cullinane—all were founders and remain primary forces in setting the directions and strategies of their companies.

As I talked with successful challengers, I tried to go beyond the daily events in the computer industry. Each day in the business press and trade journals, we read about new products and the reactions of industry pundits. Products come out with a roar, but they often fade as they battle for acceptance in the industry infrastructure. Current products announcements provide little insight into the underlying trends or long-term directions of the industry. So as I chatted with the challengers, I tried to probe beyond the current products and technologies, looking for the trends and strategies that are driving the computer industry today.

The challengers generously shared their insights and talked about their strategies. They talked about not only their successes, but also their failures and their challenges. They discussed the evolution of computing technology, computer users' attitudes, and the nature of the industry itself. They discussed not only the difficulties of competing against IBM, but also the difficulties of competing in an industry that is constantly in flux.

Most of the people I interviewed have competed directly with IBM and survived—even thrived. Some have been more successful than others. All expressed admiration for IBM, often reminding me that Big Blue remains very powerful in the computer industry. But, admiration aside, these leaders aren't caught up in the Big Blue mystique. They don't spend much time reacting to IBM moves, and they don't base their strategic plans on predictions about IBM's next moves. Instead, they are obsessed with developing their own markets, meeting the needs of their own customers, and providing innovative solutions to problems that IBM has not yet considered. They view the market in terms of opportunities—opportunities they create for themselves, independent of IBM.

As I talked with the challengers, what came across was a pattern of consistent values and strategies that they use as guidelines for competing effectively and running their businesses. The challengers share many ideas about what it takes to succeed in the ever-changing computer industry. They are all masters at anticipating change, planning their products and management so that they can weather changes within the industry. But challengers do not simply ride out the changes: they provoke them and extend them. In short, they become agents of change themselves. They do not simply respond to markets and trends, they create them.

Perhaps most important, all of the successful challengers are *sustainers*. That is, they are not one-time innovators; each has introduced a series of innovations. Not all of the challengers are as revolutionary as Cray, Digital, or Apple. But all have a history of *sustained innovation*:: they continue to introduce new products, make incremental improvements, adapt existing products to better meet the needs of customers. It takes more than technical prowess to succeed in the computer business. It takes singleness of purpose, focus, drive, and a sustained willingness to compete even when the whole world seems to forecast your doom.

❑ ❑ What's Ahead

The rest of this book is divided into three parts:

■ In Part One, I begin by describing IBM's rise to power. Then I discuss the fundamental changes in the computer industry that are now threatening IBM's position—and presenting opportunities for IBM's challengers.

■ In Part Two, I describe the strategies that the challengers are using to nibble away at IBM. In particular, I focus on six core elements of strategies for competing in today's fast-changing markets.

■ In Part Three, I look ahead at future directions for IBM, the challengers, and the computer industry as a whole. I don't actually try to predict the future. That would be a futile exercise: changes in technology-driven industries are almost always unpredictable. Rather, I offer suggestions for how to survive and succeed in an unpredictable future.

PART ONE

The World of Big Blue

> *The more complex the system, and the more "software" it requires (such as operating procedures and protocols, its management routines, its service components), and the longer it takes to implement the system, the greater the customer's anxieties and expectations. Expectations are what people buy, not things.*
>
> Ted Levitt, *The Marketing Imagination*

Growth of a Giant

■■ THE HISTORY OF IBM is, of course, a history filled with success. During the first half of the century, under the leadership of Thomas J. Watson Sr., IBM grew into a major corporate power, creating and building new markets for accounting machinery. Thomas J. Watson Jr. assumed the presidency in 1952 and built on his father's success. The younger Watson ushered IBM into the age of computers, leading the company to a position of dominance unmatched in business history.

Yet IBM's history of success is a story with an ironic twist. Many of the very strategies and practices that brought IBM to a position of dominance are now causing the company to falter. IBM's style, size, and history, long a source of corporate strength, are now a source of corporate vulnerability. Today's business world is far different from the world of the Watsons, but IBM is, in many ways, a captive of its past.

There are plenty of long books on IBM's history, so I plan to present only the sketchiest details of the company's history. In order to understand the computer industry today, it is important to have at least some sense of IBM's past. For it is in the past that we can see the seeds of IBM's current problems—and the opportunities for IBM's challengers.

□□ Sales as the Key to Success

In building IBM, Tom Watson Sr. created a model of the modern industrial company. He introduced a wide range of new management and sales techniques that were admired and imitated throughout American industry. Indeed, Watson's influence can still be seen in management practices today.

It is possible to trace many of Watson's ideas to his early work as a salesman for National Cash Register. At NCR, Watson found a mentor in chief executive John T. Patterson. From Patterson, a pioneer of modern sales methods, Watson learned many of the basic business tenets he later zealously applied to IBM.

At a time when salesmen were widely perceived as charlatans, Patterson upgraded the salesman's role at NCR and richly rewarded his salesmen for jobs well done. Patterson revolutionized sales practices at NCR. At most companies at that time, management purposely overlapped sales territories, aiming to encourage competition among salesmen. Patterson strongly rejected such in-house competition. Instead, he built team spirit by guaranteeing his salesmen individual territories. He originated the 100 Point Club, an exclusive club for NCR salesmen who brought in $30,000 worth of business in a one-year period. This club served as a model for the 100 Percent Club that Watson later introduced at IBM.

Although he was increasingly erratic toward his employees as they rose in NCR's ranks, Patterson was uncommonly generous with his general work force. Patterson theorized that well-cared-for employees would provide good service, and he put his theory to work. He provided low-cost employee dining rooms, schools, libraries, parks, and on-site showers. His experiment proved successful: NCR grew to annual sales of $22 million in 1913. Patterson's employee practices provided a model not only for Watson and IBM, but for all of American industry.

Beyond salesmanship and humanistic management practices, Watson learned valuable lessons in hard-core competition from Patterson. The NCR president loathed competition in his company's markets, and he stopped at almost nothing to annihilate it—an attitude that still runs deep in the soul of IBM. Among NCR's hardball tactics: tampering with competitors' registers, building faulty look-alike registers, setting up temporary shops near smaller competitors' stores, and grossly undercutting prices until the smaller companies were destroyed. Although Patterson and several top NCR executives were convicted of antitrust violations in 1912, the government dropped its case on appeal the following year because of a procedural flaw in the first trial.

Working to the top of NCR was a dangerous route. As Watson began to emerge as a potential successor to Patterson, he was fired, as had been many potential heirs before him. Just shy of forty years old, Watson in 1914 accepted a position as general manager with a small, unprofitable company known as the Computing-Tabulating-Recording Company. CTR had 1,200 employees and annual revenues of $4 million when Watson took it over in 1914. He vowed to make CTR a great company, and he started to apply the lessons he had learned at NCR.

Watson quickly recognized the long-term profit possibilities in the tabulating machine sector of CTR's business. The tabulating machine was an electromechanical device that manipulated information stored on stiff punch cards. Information was encoded on the cards according to the location of the holes: different patterns of holes represented different numbers or letters. Herman Hollerith had initially invented the tabulating machine to speed data analysis for the 1880 U.S. Census. Before the tabulating machine, it had taken nearly ten years to record and analyze census results. With the tabulator, the data from the 1880 census was analyzed in just two years. Hollerith started a company called Tabulating Machine Company, and was soon renting machines not only to U.S. government agencies, but to foreign governments as well. In 1910 Hollerith's company merged with another small company to form CTR.

Upon joining CTR, Watson set out to make the company the leader in the tabulating machine market. Inheriting a company with flagging employee morale, Watson immediately began implementing the humanistic management ideas he had learned from Patterson. To create a sense of community at CTR, he went so far as to institute company "sing-alongs," in which employees sang fight songs printed in official company songbooks.

Watson also set new styles and new standards for CTR salespeople. Watson prohibited drinking on sales calls, claiming that only sober men could make effective pitches. And he set the first formal dress code for CTR salespeople, a marketing technique that has survived long past Watson. Uniforms have long been perceived as indicators of high-quality professional service. Even today, most airline passengers would feel more confident with an airline pilot dressed in traditional military uniform than with one wearing jeans and a T-shirt. The IBM "uniform" marked Watson's sales team as being a cut above those from other companies.

As World War I inflated government bureaucracy, the market for tabulating machines boomed. Aggressive CTR salesmen were ready. As a result, CTR's sales doubled in three years, jumping from $4.2 million in 1914 to $8.3 million in 1917. Several years later, Watson changed the company's name to International Business Machines Corporation. According to an account by Watson's son, the name change was carefully calculated: Watson wanted to create an image of greatness.

IBM built on its success in tabulating machines to establish an overwhelming leadership position in the business machine industry. The company's success was based in part on Watson's competitive zeal and skill in building employee loyalty. But equally important was his ability to take advantage of evolving market perceptions. Before Watson joined CTR, business machines were widely perceived as a separate category from industrial machinery. In the common view, the business-machine category included products like filing cabinets, typewriters, and pens. The industrial-machinery category included larger factory machines, operated by blue-collar workers. Perceptions of these markets were well defined, even though some companies' products overlapped the two market boundaries. Hollerith's tabulators, originally designed for the U.S. Census Bureau, did not fit naturally into either category. Most important, tabulators were not seen as a new type of business machines, and thus attracted little competition from traditional business-machine companies. This perception lingered even after Hollerith placed tabulators at insurance companies and railroads, traditionally strongholds of business-machine firms such as Underwood and Burroughs.

Watson took full advantage of these outdated, but still widely held, market perceptions. CTR leased tabulators to numerous large businesses, establishing an installed base in the business-machine market largely unnoticed and certainly unhindered by significant competition. By the mid-1920s, IBM was leasing 85 percent of the tabulating machines, 87.6 percent of the sorting machines, and 81.6 percent of the punch-card machines in the United States. It also owned a whopping 85 percent of the closely linked punched-card market, selling about 3 billion cards a year. Ironically, the same kind of shifting market perception and definition that helped IBM in the 1920s is undermining the company's position today.

□ □ Creating New Markets

During the 1940s, when the United States entered World War II, Watson put IBM's facilities at the government's disposal. IBM built a special weapons production plant in Poughkeepsie, New York. At other IBM plants, employees worked overtime to produce data-processing equipment for wartime record keeping. IBM machines on mobile units followed U.S. troops to the front lines to reduce paperwork. IBM entered the war a $40 million company and emerged a $140 million company, increasing its work force by 50 percent and tripling its factory space.

Perhaps more important, the war years ushered in a new techno-logical age. During the war, computers began to displace tabulating machines in government and defense agencies. Two University of Pennsylvania scientists, J. Presper Eckert and John W. Mauchly, were the leading pioneers in the new field. Together, they built ENIAC (Electronic Numerical Integrator and Computer), generally recognized as the first electronic computer. Eckert and Mauchly developed ENIAC to make ballistic curve calculations for the Army, but they soon left the university to develop a civilian version of ENIAC.

Even though Watson had seen ENIAC, he failed to realize its enormous commercial potential. The computer market in 1950 was generally believed to be limited to a few government agencies, much the same way as the tabulator market was perceived in 1900. Watson accepted this conventional wisdom. IBM's line of electronic calcu-lators was doing so well that the IBM chief saw no need for major expansion into computers.

Watson's son, then an IBM vice-president and a member of the board, disagreed with his father and pressed for IBM to move into computer production. The young IBM heir was fascinated by the possibilities of computers in industry. He believed prospective customers simply needed educating about the tasks computers could perform. At first, Watson Jr. fought a losing battle inside the company. Watson Sr. could see no immediate demand for the new machines, so IBM's investment in computers was limited to research.

IBM's competitors, though, were not as cautious. In 1952, Remington Rand, sensing the potential of the new technology, acquired the company started by computer pioneers Eckert and Mauchly. The next year, Remington Rand introduced a computer called UNIVAC. The first model was delivered to the Census Bureau, where it replaced some of IBM's equipment.

Remington Rand's success with the Census Bureau woke the competitive instincts in Watson. He ordered IBM design teams to speed the development of the Model 701, a scientific computer that had been slumbering in its research phase. But IBM was starting from a position of weakness. The 701 was designed for laboratories, not business offices. And even though the 701 was IBM's most complex machine, it wasn't nearly as advanced as the UNIVAC. Meanwhile, Remington Rand wasn't sitting still. As IBM struggled to regroup, Rand was at work developing even more advanced UNIVACs, designed for specific business functions.

In trying to recapture its lead in the business machine industry, IBM went through some of the toughest times in its history. The company's game of catch-up consumed a huge chunk of resources, and the company entered the capital markets to sell bonds. IBM needed funds not only to finance its fledgling computer effort, but also to continue production of its current product line and to expand its international business. During this period, IBM sank deep into debt. In 1950 there was substantial doubt about IBM's viability, as the company struggled with major changes in its products and its structure.

IBM's role in the computer industry remained in doubt through the mid-1950s. But Watson Jr., who took over as president in 1952, had an unwavering vision of the importance of computers, and he pushed the company into the computer market with great urgency. As IBM battled to catch up with Remington Rand technologically, Watson took advantage of IBM's large installed base in electronic calculators. IBM began producing peripherals for its 600-series electronic calculators, making the calculators perform somewhat like computers. Many IBM customers, still skeptical of the new-fangled and expensive UNIVAC, were quite happy with the upgraded 600s. Many customers would later exchange their 600s for the new 700-line computers. IBM took a short-term loss on the upgrades, but the strategy succeeded in keeping Remington Rand's UNIVAC away from IBM's customers, preserving IBM's established position in the business market.

In his book *A Business and Its Beliefs*, Watson Jr. explained why IBM was able to recover from its slow start in computers. The company recovered, Watson wrote, because it

> *had enough cash to carry the cost of engineering, research and production. Second, we had a sales force whose knowledge of the market enabled us to tailor our machines very closely to the needs. Finally, and most important, we had good company morale. Everyone realized that this was a challenge to our leadership. We had to respond with everything we had—and we did.*

After establishing its Model 701, IBM started taking orders for the 702, even before the new computer was in production. IBM continued developing new computers at a near-frenetic pace. Before the first 702 was shipped in 1955, IBM took orders for two more models, the 704 and the 705. The 705 replaced the widely used 600-series calculators, and its price conveniently undercut Remington Rand's UNIVAC. Thus, in a concentrated effort, IBM succeeded in protecting its broad customer base as it caught up with Remington Rand technologically. In the process, it wrested the lead in computers from its most viable competitor. Although IBM would continue its battle for increased market share, it would never again slip to Number Two in the mainframe computer business.

❏ ❏ Maintaining the Lead

Other firms continued to sell computers to customers in the established markets of defense and government agencies. But in the rapidly growing office market, there was no contest. None of IBM's competitors had any experience in the office equipment environment, while IBM had been establishing its position in the office since it sold its first punch-card equipment.

Many companies developed technologies superior to IBM's, but they were never able to capitalize on the technological advantage. Several companies developed long-life vacuum tubes that were far better than those used in the first UNIVACs and the first IBM computers. And in 1948, AT&T's Bell Laboratories developed the transistor, the technology that ultimately replaced the vacuum tube. Three years later, AT&T licensed the basic patents to other companies. Philco, RCA, and General Electric quickly developed computers based on the new technology. All three companies marketed computers much more advanced than IBM's machines at the time. But in the market environment of the 1950s, superior technology was not enough to gain market share. IBM's superb sales force, and its wide recognition in the office market, proved more important than advanced technology. IBM continued to increase its lead in the market.

In the 1960s, there were eight major manufacturers of mainframe computers: IBM, Sperry Rand (UNIVAC division), Control Data, Honeywell, Burroughs, General Electric, RCA, and NCR. But IBM was so dominant that the group was commonly known as Snow White and the Seven Dwarfs.

Collectively, the Dwarfs brought innovation and competitive pricing to the computer marketplace. They also introduced the idea of "plug compatibility," developing computers that could run IBM software and peripherals that could work with IBM computers. NCR and Burroughs were the most successful of the Dwarfs, largely because both catered to a specific market niche: the banking industry. In addition, NCR enhanced its cash registers with computer-like functions for sale to the retail industry.

But none of the Dwarfs could compete against IBM across multiple markets, and none was able to shake IBM from its position of dominance. A major reason for IBM's stronghold on the mainframe market was its large and loyal installed base of customers. Once a customer signed on with IBM, it was unlikely to switch to another vendor. IBM customers made major investments in IBM applications, training, and software. Switching to a competitor would mean throwing away years of work and millions of dollars in investment. So when an IBM user needed another computer, it almost always went back to Big Blue.

New computer users also had good reason to buy from IBM. By choosing IBM computers, they gained access to all the application programs that had been developed by IBM customers over the years. In many ways, IBM's customers paid for and established IBM's value-added leverage more than IBM did. According to one estimate, IBM customers were using some $300 billion worth of software by 1986. IBM had developed only 5 percent of this software; IBM customers had developed most of the rest. This huge reserve of IBM software was very appealing to new customers. Thus, it was extraordinarily difficult for competitors to make inroads into IBM's market share.

IBM's emphasis on leasing further strengthened its relationship with customers. Large systems, from tabulators to mainframes, have always been the heart of IBM's business. From the beginning, IBM focused on leasing, not selling, these big systems. IBM's leasing agreements generally included service and training as part of the price. Thus, leasing arrangements kept IBM personnel in continual contact with customers, monitoring their needs, giving helpful advice, and guiding future computer purchase decisions.

Through these interactions with customers, IBM established a powerful image as a "partner" and a "friend." Customers perceived IBM as collaborator as much as vendor. In his book *The IBM Way*, Buck Rodgers, a former vice-president of marketing at IBM, discusses the power of this carefully cultivated image. Rodgers tells an anecdote about a sales pitch to a large bank:

> *It was a tough competitive situation. Another company had made a strong pitch for the business, and as the meeting, at the bank's headquarters, progressed, it became evident that the executives were leaning toward the other company. The bank's president was very impressed with the technological aspects of the competitor's presentation. He was obviously caught up in the bits and bytes, and I would have antagonized him had I debated over their relative value.*
>
> *I listened very attentively as he had his say, and when he was finished I responded. "I have only one thing to ask, and nothing more," I said.*

"Do you want to do business with a hardware vendor or do you want a partner?"

There was nothing tricky or clever about the question. It was honest. I believed that we had something very special to offer this account, something that went far beyond the technology of our machines—our sincere interest in their well-being.

"I want a partner," he said, after considering the question for a few moments. Then he walked over to me, extended his hand and said, "Buck, shake hands with your new partner."

This IBM-as-partner image not only won new accounts, it created intense loyalty among existing IBM customers. Users were confident that IBM would take care of them and their computer needs. In purchasing decisions, emotional security proved more powerful than price or technological superiority.

When customers did think of wandering from the IBM fold, IBM resorted to a set of scare tactics collectively abbreviated FUD—Fear, Uncertainty, Doubt. If a customer was about to shift to another brand, perhaps for cost reasons, an IBM salesperson might hint that IBM was about to introduce a breakthrough product that would make the competitor's product obsolete. Or the IBM salesperson might plant some concerns about the competitive vendor's long-term financial stability. Customers, filled with fear, uncertainty, and doubt, typically stuck with IBM. FUD did its job, keeping customers within the protective shelter of Big Blue.

The early history of Amdahl Corporation provides a classic example of FUD. Top mainframe designer Gene Amdahl quit IBM to launch his own company in 1970. His goal was to build a computer that could run the same software as IBM mainframes, but offer higher performance than existing IBM machines. By 1975 Amdahl was selling a computer four times as powerful as IBM's 360/165 for the same price, approximately $3.5 million. After two years, and sales of 55 of the Amdahl machines, IBM responded with its FUD strategy. Even though Amdahl had responded to IBM price cuts and product improvements with cuts and improvements of its own, a devastating rumor began spreading throughout the computer industry: IBM was about to introduce a new line of large computers that would compete directly with Amdahl's machines. Suddenly wary, potential buyers began leasing Amdahl machines rather than buying them outright. As a result, Amdahl barely broke even in the second quarter of 1979. IBM's FUD strategy had worked again: IBM had protected its turf.

❑❑ IBM the Invincible?

By the 1960s, IBM seemed invincible. Big Blue loomed above its competitors, its share of the mainframe market rising as high as 76 percent. In 1965, IBM owned a whopping 65 percent share of the entire computer market. Although IBM's share shrank somewhat during the 1970s, it maintained its position of dominance.

But the strategies that worked for IBM in the 1950s and 1960s and even the 1970s started to run into trouble in the 1980s. One problem was IBM's conservative, risk-averse culture. Watson Jr.'s bold move into computers was an anomaly for IBM. Since its early days, when Watson Sr. imposed a formal dress code, IBM has trained its workers to think and act in bureaucratic, institutional ways. Entrepreneurial and independent thinking have always been discouraged among the blue-suited army of IBM managers.

In general, IBM has resisted change. As long as the market was changing slowly, that was okay. When mainframes were the only game in town, the computer market was a market-share battle. The company with the most resources generally won the biggest share of the market—and IBM generally had the most resources. New technology was rarely enough to upset the balance of the market. People continued to buy from IBM even if other vendors offered mainframes with superior technology.

But IBM stuck with its mainframe mentality even as smaller, more versatile, and less expensive machines began to appear. As the pace of technological change accelerated in the 1970s, challenger companies began using technologies in new ways. Instead of fighting IBM for a share of the mainframe market, challenger companies began designing new types of computers and creating totally new markets. IBM, meanwhile, stuck with its old sales-oriented approach to marketing. It kept trying to improve its sales pitch, rather than adapting its technology and its machines to meet the evolving needs of customers.

The entrepreneurial startups of the 1980s presented an entirely new challenge for IBM. In the mainframe market, IBM had always battled against other large, slow-moving companies. But by the early 1980s there were hundreds of small companies battling against Big Blue. IBM was accustomed to battling tanks against tanks, not against a gang of guerrilla fighters. Just as the United States was ill prepared for the guerrilla warfare of Vietnam, IBM was ill prepared for the new competition from entrepreneurial challengers. IBM's size worked against it: the giant company couldn't shift directions as quickly as its new challengers.

As the challengers offered new and innovative machines that addressed more and more of the customers' real needs, customers began to question the image of IBM as partner and friend. Some customers saw the image as just that: an image. The image was not backed up with enough substance. If IBM were really such a friend, why wasn't it more responsive in adapting its products to meet customer needs? Why wasn't it more innovative in providing integrated solutions? Why didn't it advance open systems architectures? And why were all of the intriguing new technologies, solutions, and applications coming from the computer newcomers, not from IBM?

In short, the strategies that led IBM to power began to falter—and they continue to falter today. Today's computer environment is fundamentally different from the environment in which IBM rose to power. The next chapter looks more closely at exactly how and why the computer industry changed, and why challengers no longer need to be quite so afraid of Big Blue.

*Since electric energy is independent of the place or kind of
work-operation, it creates patterns of decentralization
and diversity in the work to be done.*

Marshall McLuhan, *Understanding Media*

□ 2
The Game Changes

■■ IN THE PAST DECADE, everything related to computers has changed dramatically. I'm not talking about gradual evolution, I'm talking about unrelenting radical change that has fundamentally altered working and living conditions for millions of people throughout the world. As a result, things will never be the same for IBM—or for the computer industry as a whole.

Once limited to "number crunching" in the backrooms of accounting offices and in scientific laboratories, computers now abound in just about every kind of business imaginable. Computers turn up in grocery stores, on factory floors, in hospitals, on secretaries' desks, in clothing stores, in car dealerships, and in the schools that educate our children.

Other computers, in the form of microprocessors, operate transparently, hidden from view. There are computers in today's power tools, appliances, telephones, air conditioning and heating systems. Computers are involved in just about every piece of mail we receive. Even though we don't see them, computers are making our cars more fuel-efficient, our traffic lights "smarter," and our lives more efficient and productive. Computers, both apparent and transparent, have permeated our society.

In 1970, fewer than 50,000 computers were in use. Today, more than 50,000 computers are produced *every day*. These numbers, dramatic as they may be, provide only a hint at the extent of change in the computer marketplace. Computers are being used in new ways, by new types of users. People are thinking about computers differently—and thinking about computer vendors differently.

The changes in the computer market have been brought on by a set of interrelated trends, each of which amplifies the others. These trends include:

- ■ **Electronic advances**
- ■ **Personalization of the computer**
- ■ **The rise of networks**
- ■ **Fragmentation of the market**
- ■ **A new breed of user**
- ■ **Mixing and matching**
- ■ **Strategic computing**
- ■ **Internationalization**

Radical changes in the technologies and uses of computers mean radical changes for the companies that design and sell computers. This chapter explores these trends, with an eye on the implications for IBM and its challengers.

❑❑ Electronic Advances

ENIAC, the first electronic computer, weighed 30 tons, filled a room the size of a one-bedroom house, and included 18,000 vacuum tubes. Two women punched numbers into desk calculators for a year to test whether ENIAC produced the right answer for a particular test problem. ENIAC solved the problem in an hour. In a weekend, ENIAC could do 15 million multiplications—a job that would take a person about forty years using the desk calculators prevalent at the time. Today, microprocessors small enough to fit on a thumbnail can match ENIAC's power. Modern supercomputers can do 250 million multiplications in a single second—about 172,800 times faster than ENIAC.

The elephantine computers running on vacuum-tube technology quickly became outdated as the aerospace and defense industries encouraged engineers to develop ever-smaller electronic devices. At first, engineers simply improved vacuum tubes, but then physicists developed an entirely new approach for manipulating electricity: the solid-state transistor. Transistors enabled engineers to miniaturize electronic designs, not only because transistors were smaller than vacuum tubes, but also because they generated less heat and could be placed closer together. By the late 1950s, transistors were replacing vacuum tubes in almost all applications.

Transistors quickly gave way to the next generation of technology—integrated circuits. Integrated circuits, or ICs, combine a number of transistors (or other simple electronic elements) on a single chip of silicon, making it possible to embed an entire circuit on a chip. At first there were only a few elements on a chip, but by the mid-1960s thousands of interconnected miniature transistors were being placed on tiny silicon chips. Because each chip performed a limited function, the central processors of large computers often used thousands of integrated circuits. Chips and other components were soldered onto printed circuit boards. In a way, the soldered connections served as a program.

Ted Hoff, an engineer at the microchip maker Intel, had other ideas about how to use and program chips. In 1969, just a year after the founding of Intel, Busicom, a Japanese calculator manufacturer, asked Intel to produce a design for a complex new calculator. Thinking the design too complicated to be cost-effective, Hoff proposed a dramatically new approach. He placed all the circuitry needed for a computer's central processing unit (CPU) on a single chip. Then Hoff figured out a way to store on a second chip the program needed to drive the CPU. A third chip shuffled data in and out of the CPU.

Hoff's new approach used just three chips, as opposed to the original eleven in the Busicom design. But more important was the conceptual change inherent in Hoff's design. The idea of putting an entire CPU on a single chip was a fundamental shift. The new CPU chip—known as a microprocessor—could be mass-produced and programmed for thousands of applications. Intel's 4004 microprocessor chip, measuring less than one square inch, was about as powerful as the room-sized ENIAC had been thirty years earlier.

Soon, systems designers began replacing all their hard-wired logic systems with microprocessors, substituting program code for hardware. Hoff's design was soon simplified even further, making the production of microprocessors less labor-intensive than producing transistor-based circuits. Costs fell dramatically. At an industry conference in the early 1970s, one of the panelists chided Bob Noyce, inventor of the integrated circuit and founder of Intel, that people would have to be careful with the new chips: if you weren't careful, you could drop your computer (chip) on the floor and lose it. Noyce quickly replied that it wouldn't matter. "Microcomputers will be so cheap that replacement costs will be negligible," he said.

Noyce was right. Each technical advance has decreased production costs, promoting more and more applications, so that computational power is now almost free. Indeed, the cost of computing is so low that computers (in the form of microprocessors) are now commonplace in American homes. The average American home contains 15 to 20 fractional horsepower motors. That development took almost seventy years. The microprocessor has taken less than twenty years to reach almost the same level of penetration. Microprocessors, produced in quantity for just a few dollars a chip, command processing power that would have cost $100,000 in the late 1960s.

The decreasing costs of microprocessors and other integrated cirucits have brought down the prices of all types of electronic products. Pocket calculators present a clear example. Early calculators were desktop business machines, often costing about $7,000. By the early 1970s, consumers could buy simple pocket calculators for less than $200. Today, calculators are cheap enough to give away at store openings and bank promotions.

Today, technological change is feeding back on itself. New technologies—like silicon compilers, computer-aided software engineering, and application-specific integrated circuits, to name just a few—are accelerating the pace and diversity of technological change. This rapid pace of change will continue to push down the prices of electronic products—and it will continue to wreak havoc for slow-moving corporate giants like IBM.

❏ ❏ Personalization of the Computer

The invention of the microprocessor fundamentally changed the way people thought about computers. For the first time, it was possible to build a complete computer that was cheap enough for an individual to use, yet powerful enough to do serious work. In short, the microprocessor paved the way for *personal* uses of the computer.

In the past decade, we have seen the rise of personal computers, portable computers, hand-held computers, and personal workstations. Some industry people are even talking about "personal super-computers." Supporting these personalized machines is a wealth of personalized software: software written to help individuals do their jobs better. Information power is now in the hands of the individual users who need it. The terms *user friendly* and *human interface* are just other ways of saying "personal."

Personal computers first moved into the public limelight in 1975, when *Popular Electronics* magazine featured on its cover a computer produced by a tiny startup company called Micro Instrumentation and Telemetry Systems (MITS). Using Intel's 8-bit microprocessor at its core, the MITS computer, known as the Altair, cost $395 as a kit and $621 assembled, not including peripherals like a monitor. The Altair was aimed at electronics hobbyists, who thoroughly enjoyed assembling, programming, and playing with the system.

Other companies joined MITS in offering computers to hobbyists. But the real impact of personal computers was not felt until Apple Computer introduced its Apple II computer in 1978. Apple co-founders Steve Wozniak and Steve Jobs recognized the enormous social and technological impact that personal computers would have on American businesses and consumers, and they set out to change the way people *thought* about computers.

Apple succeeded in radically changing the public perception of the computer. The computer was no longer seen as an automated power broker; it was a machine of the people. Apple redistributed the computer power of the few (big business, big government) to the many (common, ordinary people). Suddenly, computers could be understood and controlled by anyone who could afford to buy them. The mystique of the computer began to evaporate. Children began playing computer games, teenagers began writing software programs, and a growing number of businesses began buying personal computers to boost individual employee productivity.

One particularly telling article, which appeared in *Forbes* magazine in 1979, told of the growing use of personal computers in areas other than the hobbyist market. A half-page photo featured Ben Rosen, the most respected electronics analyst in the country and senior analyst at the conservative Wall Street investment banking firm, Morgan Stanley. Rosen was sitting at his Apple II, which he used as an analyst's tool. That photo gave the personal computer instant credibility in the business world. Apple's computer revolution was an overwhelming success. By 1980, just two years after the introduction of the Apple II, Apple's revenues had soared to $117 million, and the stock market valued the company at an incredible $1.78 billion.

One of the clearest, and perhaps most controversial, illustrations of Apple's continuing effort to alter the public's image of the computer was its dramatic commercial during the 1984 Superbowl. The sixty-second commercial had an Orwellian theme. An audience of shaven-headed clone people in identical dark uniforms sat transfixed as Big Brother blared from a giant movie screen: "The thugs and wreckers have been cast out. Let each and every cell rejoice!"

As Big Brother's commanding voice continued, a young woman in a jogging suit burst through the zombie-like audience and heaved a sledgehammer through the screen. As the screen shattered, and the image of Big Brother faded away, an announcer proclaimed: "On January 24, Apple will introduce Macintosh. And you'll see why 1984 won't be like *1984*."

The message was clear: the young Apple revolutionaries viewed IBM as Big Brother. For the good of the world, as well as the good of their company, they wanted to beat IBM—and they believed they could. At the time, the Apple revolutionaries were naive: their products weren't ready to challenge IBM in the business market. But Apple was right about the direction of the industry: IBM was no longer invulnerable. A huge new market was opening up, and IBM wasn't in the lead. The days of IBM domination were numbered.

❏❏ The Rise of Networks

The success of the Apple II (or of Apple as a company) was far less threatening to IBM than the attitude change that accompanied personal computers. At large companies, employees became less reliant on large mainframe computers for small-scale computer tasks. The mainframe computer began to lose its position of dominance in the computer industry, and that could only be bad news for IBM, the dominant supplier of mainframes.

The shift from mainframes had actually begun in the 1960s, when Digital Equipment introduced the first minicomputer, a scaled-down computer aimed at serving the needs of small businesses, research groups, or corporate departments. In the 1970s, Digital and fellow minicomputer manufacturers like Data General and Hewlett-Packard grew rapidly, as computer users distributed some of their computing tasks away from centralized mainframes to departmental minicomputers.

The personal computer, and the new technologies that made it possible, sharply accelerated the shift away from mainframes. There are now more than 15 million personal computers in American offices. Although computers are increasingly used one-to-one, by individuals, that does not mean that computers are becoming "stand-alone" units. On the contrary, more and more computers are being connected into networks. Using these networks, users of small computers can communicate, share resources, and work together on tasks. In many cases, the network is becoming more important than the computers themselves. By linking their personal computers into networks, users can share resources and amortize the investment in the information stored on the computers—the *raison d'être* for buying the computers in the first place. In short, *networking* is just another word for *sharing*.

In many applications, networks of small computers are filling roles once filled by general-purpose mainframes. At American Savings and Loan Association, for example, the trading of mortgage-based securities is handled by a $100,000 network of personal computers. According to an article in *Business Week,* a similar mainframe-based system would have cost at least $500,000, and maybe as much as $3 million.

Thus, success in the computer industry now depends largely on a company's ability to provide machines that can be linked together over computer networks. This trend cuts against IBM in two ways. Not only does it open up new market areas where IBM has no historic position, it also allows other companies to link into the IBM-installed base.

Many IBM challengers have created programs and networks to

enable their machines to work with the IBM equipment already in place. As a result, computer users are routinely looking for computers outside of IBM's product line to meet their needs. Mainframes are still extremely valuable in business and government environments, and IBM still holds the lion's share of that market. But the market itself is growing more slowly as personal computers and minicomputers eat away at what was once exclusive mainframe territory.

The move toward distributed network solutions has created opportunities for many new companies. Sun Microsystems and Apollo Computer found—or, more accurately, created—a niche that didn't exist five years ago. The two companies produce technical workstations: high-performance personal computers designed primarily for use by scientists and engineers. These workstations offer about four times the processing power and ten times the memory of the average personal computer. In most situations, networks of these workstations provide much higher performance than mainframe-based systems. In the mainframe approach, each user has a terminal linked to the central mainframe computer. The problem is that the central computer must service all users, giving a little time to one user, then a little time to another. In other words, all users compete with one another for precious computer time.

Engineering workstations take a very different approach. Individual scientists and engineers have their own private computers. Each workstation, by itself, is less powerful than a mainframe computer. But individual users have complete control over their own computers; there is no more competing for time. Thus, productivity can rise dramatically. And users can still share data, programs, and designs with one another, since workstations can be linked together through a network. Customers are rapidly recognizing the advantages of networked workstations. Just seven years after the introduction of the first technical workstations in 1980, the market had already grown to $2.5 billion in annual sales.

Digital has been a chief driving force—and beneficiary—of the trend toward distributed networks. Digital's strategy has been to build networks first and worry about peripherals (even the computers themselves) later. The strategy has brought Digital extraordinary success. While the rest of the industry slumped in 1985, Digital surged ahead. And in its 1987 fiscal year, Digital's earnings jumped 84 percent to $1.1 billion, on sales of $9.3 billion.

Digital's success in creating network solutions has had a strong impact on customers: it has broadcast the message that IBM is lagging behind other companies in serving the needs of major customers, particularly in departmental computing and engineering/scientific computing. As customers see increasing numbers of effective non-IBM computer solutions, they are becoming more assertive, more willing to stray from the IBM fold. According to one estimate, Digital has won twenty-five new customers in the insurance industry, formerly a "true Blue" IBM stronghold.

"It has really just hurt IBM terribly in terms of their image at the top," says Dave Martin, president of National Advanced Systems. "The recent onslaught of articles on DEC is very visible to upper management. They all say DEC does a better job than IBM. What has that done? It has said more than one company can do the best job for certain applications in information processing. In this case, DEC does it best."

John Sculley describes the evolution of the computer market from a unique perspective: he was a major IBM customer as president of Pepsi from 1977 until 1983, and is now a major IBM competitor as chief executive at Apple. "The world I came from was one where IBM was the oracle. None of us really questioned whether there was an alternative point of view because IBM had been so successful when it introduced the IBM 360 in 1964. That demonstrated that productivity could be defined at an institutional level by taking complex tasks and putting them on a mainframe computer and getting things to happen a lot faster."

Sculley continues: "That paradigm worked extremely well through the rest of the 1960s and all through the 1970s. But what we've seen at Apple and in the industry over the last few years is that the epicenter has shifted from the mainframe, which is IBM's center of gravity, to the network. And IBM really wasn't in control of that change, even though they dominated the attention of corporate America with their concept of what computing was all about."

Sculley believes that much of the credit for shifting the epicenter to the network should go to Digital. "Digital alone didn't shift it, however. The shift was the result of a logical progression of technology where processing power was becoming less expensive and it was therefore very possible to process the data wherever you chose. But the shift only made sense if you could connect computers together. Digital was the one that recognized what the implications of the reduced cost of processing power were, and they realized that you had to have a very rich intelligent network, and you had to have tremendous consistency within your own product line if you were going to be successful."

Edson de Castro, founder and long-time chief executive of mini-computer maker Data General, believes that the shift toward distributed computing is "probably as threatening a transition as IBM has ever seen. It is really a transition in the way people like to use computers. We talked about the demise of the mainframe for years and years. But it is really happening now. People are not putting new applications on mainframes. They'll probably run the applications they've already got on mainframes forever, but in terms of new applications, people are really looking for more distributed networks.

"If you look at the state of the industry right now," says de Castro, "I think the results of that are pretty apparent. Basically, we are in a time of slow demand. Statistics show that IBM has always improved its market share during periods of slow demand in the past, basically because customers are very risk-averse. They looked at IBM as being a *safe choice*. But strangely enough, right now, DEC is taking market share away from IBM. The reason, in my mind, is that IBM has not had the solution the customer wants, which is a distributed network approach to solving its problems."

❏❏ Fragmentation of the Market

The onslaught of new technology, and the shift toward distributed networks, has splintered the computer industry into many market slices. No longer is the mainframe king of the industry. The computer industry is, in fact, an industry composed of many subindustries.

In this environment, the role of the general-purpose mainframe is gradually shrinking. New computer vendors are offering—and new customers are demanding—more specialized solutions to particular computing needs. There are not only mainframes and mini-computers and personal computers, but also workstations, super-computers, even superminicomputers and minisupercomputers.

Beyond new categories of computers, there are also new categories of computer applications. Indeed, the categories of users and applications for the computer are also fragmenting. Within each new category of computers, there are specialized machines and software packages aimed at particular types of applications. There are now software packages to support everything from molecular modeling to aircraft flight simulation, from credit-card verification to grocery store inventory and ordering control. The list goes on and on, and new niches continue to appear. Each of these new categories offers new risks for IBM—and new opportunities for IBM challengers.

The challenges to the general-purpose mainframe computer come from all directions. Some new machines, like engineering workstations from Sun and Apollo, challenge the mainframe from below. Small specialized machines do part of the mainframe's tasks at a fraction of the cost. Because their designs are tailored specifically to the needs of a relatively narrow customer base, small specialized machines often operate more effectively than an IBM mainframe.

Other companies are challenging the mainframe from above, producing souped-up scientific computers or so-called *supercomputers*. These machines skim high-end users off the top of IBM's customer base. Cray Research, for example, prides itself in providing "the fastest computers in the world." Cray sells its supercomputers to scientists and engineers for lofty prices ranging from $5 million to $15 million each. Demand is so high that the backlog typically stretches to two years.

Fujitsu and Hitachi have joined Cray in the supercomputer market, offering machines with an added feature: their supercomputers can run IBM software. Cray, meanwhile, has programmed its machines to operate on the popular Unix operating system, so that its supercomputers can now be linked to smaller machines from other manufacturers. Analysts estimate that supercomputer market revenues will reach $1 billion before 1990, cutting a chunk out of the high-end segment of IBM's mainframe market.

Even as the supercomputer steals sales from the mainframe market, another new subindustry is challenging the supercomputer. In the early 1980s, Robert Paluck and Steven Wallach noticed the need for scientific machines more powerful than high-end minicomputers (sometimes known *superminis*), but less expensive than supercomputers. So Paluck and Wallach started a new company called Convex Computer, and the market for "minisupercomputers" was born.

Convex neatly filled the gap between superminis and supercomputers. At the time, the fastest superminis cost about $400,000 and ran about 200,000 floating-point operations (FLOPS) a second. A Cray 2 supercomputer could run at about 1.5 billion FLOPS, but it cost a hefty $17 million. In April 1985, Convex began selling computers that ran at one-quarter the speed of a Cray, at a cost of only $500,000. And by 1986, Convex was already offering an improved second-generation product line that overlapped the performance of the Cray machines.

Power-hungry computer users have flocked to buy minisupercomputers. In the first three years of its existence, the minisupercomputer market reached $176 million in sales. New competitors have crowded into the market, offering at least three different architectures to get the job done. The market, which didn't even exist before 1981, is expected to reach $1.2 billion by 1992—without any participation from IBM.

As new niche markets and subindustries continue to emerge with regularity, the number of competitors in the computer industry continues to expand. The tools for designing new computers— engineering workstations, design software, computer chips—are readily available. Just about anyone with an idea can design a computer, and at times it seems as if everyone is doing just that. Of course, wrapping a business around a technology is not easy. But everyone has a chance to enter the game.

In this new environment, there is a premium on nimbleness and flexibility. Companies that can spot new niches and react quickly are the most likely to succeed. In many cases, small startups are at an advantage. New companies, with computers based on new advanced architectures, are not burdened with old baggage. They don't need to worry about compatibility with an installed base of obsolete machines. Instead, they can quickly and efficiently design computers to meet current needs.

Thus, IBM's monolithic role in computers is eroding, as smaller, less expensive machines replace Big Blue's general-purpose main-

1987 VENDOR MARKET SHARE

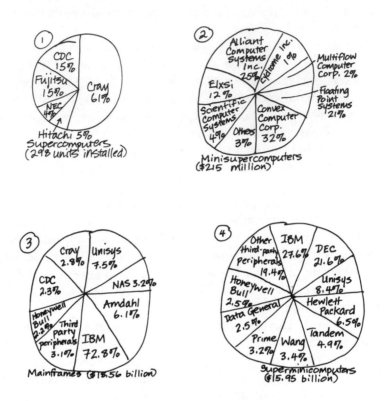

① CDC 15%
Fujitsu 15%
Cray 61%
NEC 4%
Hitachi 5%
Supercomputers
(298 units installed)

② Alliant Computer Systems Inc. 25%
Cyclone Inc. 1%
Multiflow Computer Corp. 2%
Elxsi 12%
Floating Point Systems 21%
Scientific Computer Systems 4%
Convex Computer Corp. 32%
Others 3%
Minisupercomputers
($215 million)

③ Cray 2.8%
Unisys 7.5%
CDC 2.3%
NAS 3.2%
Amdahl 6.1%
Honeywell Bull 2.2%
Third party peripherals 3.1%
IBM 72.8%
Mainframes ($18.56 billion)

④ Other third-party peripherals 19.4%
IBM 27.6%
DEC 21.6%
Honeywell Bull 2.5%
Unisys 8.4%
Data General 2.5%
Hewlett Packard 6.5%
Prime 3.2%
Wang 3.4%
Tandem 4.9%
Superminicomputers
($15.95 billion)

frames. In today's computer industry, bigger isn't necessarily better—not for computers, and not for companies. No one company can dominate all of the new market segments. And customers are no longer willing to wait for IBM. If they want high-powered workstations, they turn to Sun. If they want a machine in the small supercomputer range, they turn to Convex.

Compaq's Rod Canion explains: "IBM is a well-managed company, but it may not be possible to manage that big and diverse of an organization well enough to stay in the lead on all the fronts of the computer market. If this is a war for IBM, the number of fronts they are up against has multiplied tremendously.

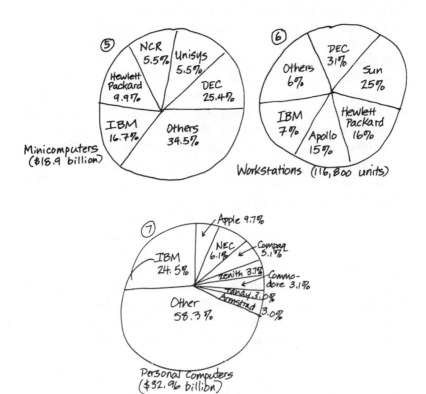

⑤ Minicomputers ($18.9 billion)
- NCR 5.5%
- Unisys 5.5%
- Hewlett Packard 9.9%
- DEC 25.4%
- IBM 16.7%
- Others 34.5%

⑥ Workstations (116,800 units)
- DEC 31%
- Others 6%
- Sun 25%
- IBM 7%
- Apollo 15%
- Hewlett Packard 16%

⑦ Personal Computers ($32.96 billion)
- Apple 9.7%
- NEC 6.1%
- Compaq 5.1%
- Zenith 3.7%
- Commodore 3.1%
- Tandy 3.0%
- Amstrad 3.0%
- IBM 24.5%
- Other 58.3%

Sources for data: (1) Gartner Group, Inc.; (2) Hambrecht & Quist, Inc.; (3) and (4) International Data Corporation (IDC); (5) Gartner Group, Inc.; (6) Gartner Group, Inc., and Regis McKenna estimates; and (7) Dataquest.

"It isn't like in the old days," says Canion. "I'm sure they [at IBM] say that all the time: this is not like the old days."

❏❏ A New Breed of User

As if changes in technology weren't enough, IBM and other computer companies must cope with an equally important change: a new breed of computer user.

The longer computers are part of our lives, the more comfortable we become with their presence. Practically everybody interacts with computers these days, and it is safe to say that more than 90 percent of the people using computers today were not using computers ten years ago. Users range from seven-year-olds playing games, to business executives manipulating spreadsheets, to aeronautical engineers designing airplane wings. Information about computers is everywhere. On the whole, computer users are becoming more confident, more knowledgeable, and more demanding.

It hasn't always been that way. In the 1950s, people were awed by the size and complexity of massive mainframe computers. Most people had only seen pictures of the mystical machines; they had never actually touched one. At computer installations, the machines were programmed and operated by teams of technical experts. In the popular mind, computers were linked to an image of white-coated technical wizards.

In this environment, IBM was able to establish a very special type of relationship with its customers. IBM eagerly stepped into the role of computer adviser, providing customers with technicians, trainers, and repairmen. Computer users relied on their IBM advisers not only for advice on new computer applications, but also for advice on which machines to buy for those applications. For their part, IBM salesmen studied and thoroughly understood the companies and industries they were selling to. From both sides, the interaction was more like a partnership than a vendor–customer relationship.

Perhaps the most important IBM selling point was its support of its products. IBM customers knew their machines would always be up and running. It wasn't that the IBM machines were significantly more reliable than competitive brands; the key point was that IBM's support was ever-present. IBM repairmen were always available within hours—sometimes minutes—of a breakdown. Managers of corporate information systems knew that they would never be embarrassed by long computer "downtime."

But time and technology have changed users' attitudes about computing and, consequently, about the indispensability of IBM's services. As technology has become more sophisticated, it has become simpler to use. Although this might seem paradoxical, it is not: with more powerful technology at their disposal, computer designers can pay more attention to ease-of-use issues. Thus, users do not need as much hand-holding as they once did. The increasing confidence among computer users has been a major force in loosening IBM's hold on the hearts and minds of its customers.

Dave Martin, president of National Advanced Systems and a former director of marketing at IBM, recalls how IBM used to teach customers everything they needed to know about computers. "Twenty years ago, [IBM] did everything—including educating the customer on what A-B-C inventory levels meant," explains Martin. "IBM used to have a thing called application school. It was all based around something called the Ideal Milk Bucket Company. They literally took you in as an employee and trained you on what a manufacturing environment is—on how computers can be applied.

"Now, it's just the opposite," says Martin. "The vendors don't have a prayer to do that. Customers now are more sophisticated about what their information-processing needs are and how to fill them. That's the biggest change of all. Once the user says, 'I know my needs and I know how to fill them better than the vendor does,' then automatically that's going to reduce vendor control. That's IBM's single biggest loss—more than any product."

Convex co-founder Steve Wallach thinks that the trend toward user sophistication began when general engineers started using computers in their work. Engineers resisted computer vendors who tried to dictate their choices. Wallach recalls an incident from his days as a young engineer. He was approached by an overbearing computer salesman who insisted that his company's wares were best for the job. Wallach angrily replied: "I'm an engineer, I make my own decisions. You tell me you make my decisions, and you won't get the time of day. If I buy your machine, it's because *I* want to buy it!"

Although most early computers were used as administrative tools, engineers had other ideas: they began using computers as design tools. Wallach, for instance, used one of his first computers to design another computer. Engineers needed to be innovative in what they were doing; the success of their projects often depended on how they used their computers. Gradually, says Wallach, decisions about computer purchases became decentralized, and engineers began making more and more purchase decisions.

As a result, the phenomenon of the "IBM alternative" began to develop in the mid-1970s. In the scientific and technical communities, computer users felt competent enough to choose technological superiority over the IBM brand name. Scientists and engineers could appreciate the advantages of advanced computers offered by companies like Cray Research and Digital.

Digital president Ken Olsen is amazed at the progressive sophistication of his computer customers. "I never dreamed that people would be so competent now," he says. "We believed that children would learn on a computer. We knew that children could use computers to play games and do other things. But really, we had no idea that so many people would learn about computers. And we had no idea that customers could build such complex systems."

Olsen recalls one time when he underestimated customer competence. Digital was about to introduce a new networking product. "When I first saw our clustering system for tying a bunch of large machines together, I said: 'This won't sell, it's too complicated.' But the day it was done, we had more orders than we could deliver. The customers knew exactly what they wanted, how to hook it up, how to use it. The ability of the customer—an individual or a large organization—to do complex things, always surprises me."

This growing customer confidence has undercut IBM's traditional FUD (Fear, Uncertainty, Doubt) strategy. IBM can no longer scare customers away from the competition. As customers have become more confident and competent, they have become more willing to try alternatives to IBM. Companies like Amdahl provided proof of reliable, economical alternatives to IBM. Max Hopper, vice-president of information systems at American Airlines, believes that IBM's influence over customers has declined as the number of competent computer vendors has risen. "I think IBM has lost the power. Its marketing approach back in the '60s and early '70s was to overpower [competitors] through its association with senior management. I think that's largely dissipated. Quite honestly, we have an awful lot of vendors today that can offer a solution."

In the 1980s, fears associated with computers have continued to fade. As computers become easier to use and better designed for specific tasks, user confidence levels continue to rise. Forays into new technological territory now seem exciting rather than frightening.

The personal computer has had a particularly strong influence. With the introduction of personal computers, computer technology moved into the hands (and minds) of millions of users. Every major metropolitan newspaper now has a computer section or column. *Time* and *Newsweek* have computer editors. Children learn about computers in school. In short, the world is becoming more computer-literate, and that literacy has changed the perception of corporate America.

The rapid acceptance of IBM PC clones (that is, machines that run the same software as IBM PCs) graphically illustrates the increase in user confidence. Many industry analysts believed the psychological influence of the IBM image would drive all personal computer users to choose Big Blue. Instead, as computer users have become more familiar with small computers, they have gained enough confidence to move away from the established brand. In many cases, they are choosing functionality and price over brand name.

There is a similar trend at major companies: a new breed of manager is taking over control of management information systems (MIS or IT, information technology). The new MIS bosses are less likely to take their cues from IBM. They know that computers are at the heart of what it takes for companies to compete and succeed. They are no longer content to have the computer giant dictate what they should buy and how they should be structured. They want products and services that are unique, tailored to their needs.

IT managers are playing increasingly important roles within corporate hierarchies. They are emerging from glass-enclosed computer rooms to become key players in the strategic direction of the business. "They're a broader gauged people," says Joe Zemke, president of Amdahl. "Broader gauged and more experienced. They've been around longer, they're less technically oriented and more strategically oriented. I think that the technology is being used differently. American Airlines, for instance, is using technology as a profit center, as a way to break into new businesses. That's a lot different than using computers just to run the back room, or just to run the payrolls."

The new IT managers are more intelligent and demanding than ever before. "The major accounts know more about what they are looking for and how to assess things," says Computer Consoles president John Cunningham. "They won't accept a line of B.S. from the vendor. IBM always delivered them a large amount of B.S. Now customers are much more literate, much more intelligent buyers."

The computer industry, then, is shifting to what management guru Tom Peters calls a "customer-driven environment"—an environment in which users are regaining control over their computer future. Computer customers of the 1980s want it all. They want stability and assurance from computer vendors, but they also want be sure that they have the best products to do the job—and at the most reasonable price. Some call it the age of "end-user computing." End users now call the shots. The computer industry, once driven by technology, is increasingly driven by the applications that end users want.

Data General's de Castro believes that life for IBM challengers improves as users take control of their computer decisions. "I think the opportunity is always greatest when consumers' tastes are changing. When the consumer keeps buying the same thing he has been buying for years, and gets accustomed to doing business with somebody, he tends to get locked in. The customer is trained."

Today's customers are certainly not "trained." They are knowledgeable, demanding, and free-thinking. That's good news for challenger companies, and bad news for Big Blue.

❑ ❑ Mixing and Matching

In the past, computer users generally bought all of their computing equipment from a single vendor. In many cases, computers and peripherals from different vendors couldn't work together. And even if they could, users didn't have the expertise to make it happen. So users had very little flexibility. After a customer bought a computer from a particular vendor, it typically continued buying from the same vendor, even if the vendor had obvious weaknesses in parts of its line.

Today's computer customers are no longer willing to make such compromises. As customers become more knowledgeable and demanding, they are no longer willing to buy all of their equipment from a single vendor. They want their systems tailored to their own individualized needs. That might mean a mainframe from one vendor, minicomputers from another, personal computers from a third, and printers from a fourth. In short, customers want the best systems possible, no matter which brand names are on the boxes.

In some cases, customers are learning to put together such custom systems on their own. But more often, they are turning to a new breed of computer company: the *systems integrator*. Systems integrators do not actually make computers. They mix and match computers from other vendors, putting together individualized systems to meet the needs of particular users.

Increasingly, systems integrators are becoming the prime vendors to end users. If a bank needs to buy automatic teller machines (ATMs), PC's, and teleterminals, it will have a hard time finding a company with all the necessary pieces. And it is unlikely that the bank will have the expertise to put the system together itself. So it makes sense for the bank to turn to a systems integrator that can provide a turnkey solution to meet all of the bank's information needs.

The emergence of networking standards is accelerating the trend of systems integration. Communications used to be a major obstacle in multivendor systems: computers from different vendors couldn't talk to one another. But new networking standards allow systems integrators to build networks with computers from many different vendors, focusing on performance and user needs, without sacrificing communication capabilities.

Although most systems integrators are quite small, they are offering tough competition to IBM. Both are going after the same customers, and in many cases the systems integrators are winning. As Dave Martin of National Advanced Systems explains: "IBM is talking to that same customer, saying, 'Don't go with a systems integrator like Computer Sciences, go with me and I can take care of *all* your needs. Buy an all-IBM solution!' " But why should the customer go with IBM? The systems integrator can offer the best equipment from a dozen different vendors.

The systems integration trend is clearly a big plus for IBM challengers like Apple and Tandem. These companies do not offer as broad a product line as IBM, so they alone can not meet all the needs of major computer users. Moreover, challenger companies can support only a limited number of application areas. Systems integration leverages the capabilities of challenger companies. Through systems integrators, challengers can reach customers that they never would have reached by themselves.

For the challengers, then, systems integration offers a great

opportunity for expanded markets. For IBM, it means a potential loss
of account control.

❏❏ Strategic Computing

Until recently, most business computers were hidden in the back
office, grinding away at numerical calculations, processing payrolls,
and the like. Increasingly, though, computers are changing business
in more fundamental ways. Rather than simply supporting and
speeding up old ways of doing business, computers are opening up
new ways of doing business. The computer has become a strategic
resource, a competitive weapon, a vehicle for new services, a
knowledge transaction system. In short, the computer has become
the key to corporate competitiveness.

In offices, computers are providing managers with more reliable
and up-to-date information, letting them spot trends—and act on
them—more quickly than ever before. In the factory, computers are
not merely reducing labor costs. They are helping companies produce a greater variety of products, and get the new products to market
more quickly.

Companies are using computers strategically, where they once
used them simply as support services. When a company looks for a
computer vendor today, it comes with a different list of requirements
and a different mindset. In many cases, companies are buying
computers to improve information flow within the company, not to
replace clerks in the accounting department.

In a growing number of industries, these strategic uses of computers are critical to survival. Major computer users, including banks
and retail stores, airlines and manufacturers, are facing highly
competitive environments. Companies in these industries need new
ways to gain an edge on the competition. They face a stark choice:
they must innovate or vanish.

Strategic uses of computers can be found among companies of all
sizes and in all industries. Otis Elevator, for example, used a system
from Cullinet to create an on-line database of every Otis elevator in
Canada and the United States. "So when you call up and you've got

a problem with something, they know exactly what the track record will be, they know where the repair person is, how long it takes to get them, and they can do all kinds of follow-up," explains John Cullinane, founder of Cullinet. "That gives Otis the competitive advantage in selling elevators to people."

In some cases, computers can help companies move into entirely new businesses. That is happening now to Mobil Oil, the first petroleum company to enhance its gas pumps to accept debit/credit cards. As part of this enhancement, Mobil electronically linked all of the gas pumps to a central computer, using technology from Tandem Computer. But once the gas pumps were linked together, they were no longer simply gas pumps. They became computer terminals, useful for many different applications. "What is a gas pump now?" asks Tandem's Jim Treybig. "It can be anything! A gas pump can be a store: it can compete with 7-Elevens. It can be a place where you send a telex. It can be an electronic post office. You could think of any kind of thing. They could even move into banking."

One of the first—and still one of the best—examples of strategic computing is the Sabre reservation system at American Airlines. The Sabre system, developed in the early 1960s, was the first computerized reservation system to provide flight information to travel agents. The system significantly reduced booking time, and was enthusiastically adopted by travel agents. Because Sabre listed American flights first, and travel agents in a hurry tended to choose the flights they found the fastest, the system led to a big jump in American's market share. According to one estimate, Sabre generated $116 million in profits for American during its first year in operation.

American had always been a leader in uses of new technology. Before computers, American was one of the first airlines to use punch cards to keep track of passenger information. But as the introduction of jet aircraft transformed airline business in the early 1960s, American recognized that the old methods would no longer suffice. American's chairman at the time, C. R. Smith, happened to talk with an IBM executive aboard a plane one day, and he became convinced: computers were the answer.

Smith spearheaded a bold computerization effort. At a time when aircraft cost less than $2 million, American invested about $40 million in a new computer-based system—the first commercial on-line real-time system. It was a huge investment, and it required a major restructuring within the company. "It changed totally the way we were structured internally; it changed the way we did business. That wasn't easy," recalls American vice-president Max Hopper. But the risk paid off. "The customer service approach was very much our imperative. We recognized the need and acted," says Hopper.

Hopper believes that strategic computing initiatives, unlike previous uses of computers, require strong support and leadership from top management. He credits Smith with providing the vision at American. "Smith was willing to put his commitment and senior management commitment behind the new system," says Hopper. "He took a huge risk and at the same time recognized that it was going to change the way the company did business. He was willing to accept that. I think that most companies, most businesses today, don't have that. Very few companies have chosen to use information technology to achieve a competitive advantage. They may be willing to go out and automate a particular simple process. I've seen instance after instance where this product line tried to put something in place to address its specific needs, but without looking at the total business needs."

Strategic applications of computers present great opportunities not only for computer users but also for innovative computer vendors as well. By helping customers define and implement strategic computing initiatives, challengers can gain access to accounts that formerly dealt only with IBM. Tandem's work with Mobil is a notable example. "We had never sold a computer to a petroleum company," notes Tandem's Treybig. "But once Mobil had committed and they started deploying debit/credit cards at the gas pump, then the whole industry had to follow. Competition is so tough now that you just can't wait. So, all of a sudden, we had seven petroleum companies as customers."

❏❏ Internationalization

Every country, industrialized as well as newly industrialized, views the computer as a means for improving its standard of living in the next century. They all want to participate in the computer age in two ways: using computer technology and producing computer equipment.

Computer networks and applications are breaking down the divisions between traditional national markets. National markets are turning into global markets as information flows freely across national boundaries. Despite all of the talk of protectionism, computer-based services are making world markets more entangled than ever before.

Consider what is happening to stock markets. There are roughly twenty major stock markets around the world. Brokerage firms with international computer networks have access to all of these markets, so they can buy and sell stocks twenty-four hours a day. They can trade stocks in Japan, London, or New York. So, for the first time, stock markets are competing with each other across national boundaries. And, in a way, they are competing with brokerage firms as well. Someday, brokerage firms might not even go through the markets. They might move funds around the world themselves, matching buy orders in one country with sell orders halfway around the globe.

Banks are playing a similar role. Tandem's Jim Treybig explains: "We have a very large number of big banks bringing up their international systems. Let's take the Bank of Tokyo. First, they are bringing up a worldwide network to move images. But they can move anything. What you call a telex is no different than bringing up a copy of something. The banks own brokerage firms, so they tie into every country. With this international delivery system around the world, they too can compete in the brokerage business, because in many countries banks *are* brokerage firms. So the big trend is not just twenty-four-hour trading, it is the movement into a world market."

The second aspect of internationalization is international competition within the computer industry. International competitors continue to seek entry into the mainstream of the computer business. Japanese companies, of course, have already become a dominant force in the semiconductor industry, controlling nearly the entire market for certain types of memory chips. Similarly, Japan has built a dominant position in the consumer electronics industry. Now Japan is starting to make inroads into the computer industry. More than a dozen Japanese companies are expected to introduce workstations in the next two years.

But Japan is not the only international competitor in the computer industry. We are seeing a resurgence of European computer companies like Olivetti and Siemens and Bull. Industrialized countries like France, the United Kingdom, Germany, Italy, Sweden, Korea, Taiwan, and Brazil all want to build computers within their own borders.

At the same time, newly emerging countries see a native computer industry as an economic necessity for improving standard of living. A brand of technological nationalism is developing, with governments changing laws to support the local computer industry. Increasingly, when U.S. companies try to sell machines overseas, they are up against home-grown competitors. And as the foreign competitors grow stronger, they are beginning to compete in the U.S. market as well. That's bad news for IBM: almost all of the new international competitors want to beat up on IBM.

❑❑ A Decade of Change

Never in the history of business has an industry gone through changes as radical as those that have transformed the computer industry during the past decade. This chapter discussed eight of the changes that have shaken (and continue to shake) the foundations of the computer industry. In the next chapter, I examine how these changes have affected IBM, transforming the once-indomitable computer giant into a vulnerable competitor in the computer market.

It probably will not come as a surprise that less innovating companies are dominated by tall hierarchies, and that honoring the chain of command is a value.

Rosabeth Moss Kanter, *The Change Masters*

□ 3
Big Blues for
Big Blue

■ ■ IT WASN'T LONG AGO that some "experts" were proclaiming IBM invincible. As recently as 1985, the *New York Times* ran an article under the following headline:

The Daunting Power of IBM
It Keeps Growing Stronger

According to the article, some industry observers were worried that competition with IBM was becoming "virtually impossible." The article quoted the general manager at one Texas utility company: "The biggest concern is that, in the future, there might not be any effective competition to IBM."

Indeed, IBM seemed to be in a great position. Of the top 100 companies in the computer business, IBM controlled 41 percent of the revenues and 69 percent of the profits. Moreover, IBM seemed to be moving into a dominant position in the hot personal computer market. A *Business Week* cover story in 1983, proclaimed that the personal-computer battle was over—and that IBM had won.

These media proclamations seem rather foolish today, and they will seem even more foolish in the future. IBM is far from invincible—and it is becoming more vulnerable all the time. The dramatic changes in the computer industry, outlined in Chapter 2, are all playing against IBM's strengths. Size and resources are no longer the determinants of success in the computer business. Instead, there is a high premium on innovation, flexibility, and adaptability—all IBM weaknesses.

IBM was likely to run into some problems regardless of what it did: in today's environment, large size automatically brings some problems. But IBM has made matters worse for itself. Many elements of IBM's corporate culture, such as its emphasis on fighting for market share rather than opening new markets, are not well suited to today's fast-changing markets. And IBM's culture, entrenched after years of success, is exceedingly difficult to change.

The first chinks in IBM's armor began to appear more than a decade ago. Companies like Amdahl and National Advanced Systems began successfully selling "plug-compatible computers"—computers that used the same software and peripherals as IBM computers, but sold for far less than comparable IBM machines. Users could switch to non-IBM computers, but keep using the same software and peripherals that they had been using for years. At first, most computer users were willing to pay extra for the IBM name and reputation. But gradually users began to wonder why they should pay more. The plug-compatible computers offered lower prices, higher performance, and comparable reliability. As users gained more confidence, they began to shift away from Big Blue.

The rise of the microprocessor has caused further problems for IBM. The microprocessor has spawned new categories of computers and a wide range of new applications. Innovative startup companies like Apple and Sun have used the new technology to create entire new markets. These innovative startups have set the pace in the development of personal computers and personal workstations. While IBM continued to focus on big general-purpose computers, new companies targeted specific new applications, attracting millions of new customers to the computer market.

IBM's efforts to fight back in these new markets have generally faltered. At first, IBM's personal computer was a rousing success. Within a year of its introduction, the IBM PC was the industry standard. But a flock of clone-makers has chipped away at the IBM lead. Just as Amdahl offered an alternative in the mainframe market, the new clone makers offer IBM-compatible personal computers with higher performance and lower prices than IBM. Today, clone-makers own a bigger slice of the PC pie than IBM. According to Dataquest, IBM's share of the PC market fell from 68 percent in 1984 to just 37 percent in 1987. The IBM PC is still a standard, but it is no longer IBM's exclusive standard.

IBM has also run into problems in the market for midsized computers. Digital Equipment's VAX computer has stolen thousands of sales from IBM. In 1986, IBM's revenues for midsized systems actually dropped, while Digital's grew by 33 percent. The

VAX had clearly become the machine of choice in the midrange category. IBM tried to respond with its 9370 computer—known as its "VAX killer"—but the VAX killer was anything but lethal. A lack of software slowed down the initial sales effort, and the computer has never seriously threatened Digital's VAX.

At all levels, IBM has been hampered by the incompatibility of its machines. Today's customers want to link their computers together into networks, and companies like Digital and Sun have met this need: their computers connect together and share information with relative ease. But IBM's computers are a networking mess. IBM has many different families of computers, and the families are not on speaking terms.

All these problems plunged IBM into its worst tailspin since the Depression. In 1986–1987, Big Blue suffered five consecutive quarters of profit declines. Since then, IBM has recovered somewhat. But the trend is clear: IBM will never again be as dominant as it once was.

In this chapter, I trace the reasons that the changes in the computer industry have caused so much trouble for IBM. Why has the competition been able to make inroads against IBM, and why is Big Blue hamstrung in its efforts to fight back? In particular, I focus on four sources of IBM's vulnerability:

- IBM's bureaucracy
- IBM's installed technology
- IBM's incompatible products
- IBM's arrogance of success

❏❏ Problem #1: A Mind-Boggling Bureaucracy

Size can be an advantage for a company. For one thing, large companies can afford to integrate vertically, producing components and subsystems in-house. By designing and producing their own components, companies can more easily add proprietary features to

their systems. Large companies can also take advantage of economies of scale: the more units they produce, the lower the unit cost.

But size also has its disadvantages. As a company grows, so does
the risk of bureaucracy. This story is repeated over and over in the
business world. A growing company starts to lose control of its
sprawling organization, so it adds extra levels of management to
coordinate and control the flow of information. But before long, the
company is strangling on its own paperwork. The creativity and
flexibility that led to the company's initial success are squeezed out
of the organization.

IBM has fallen into this bureaucratic trap. As IBM grew into a
giant company in the 1960s and 1970s, it developed a huge, self-
sufficient bureaucratic organization. Even IBM's divisions became
multibillion-dollar operations, significantly larger than the company's
competitors. For a while, this bureaucracy was actually somewhat of
an advantage. IBM's well-staffed organization represented stability,
and computer customers were reassured by stability. The organization was slow-moving and somewhat rigid, but in the environment of
the 1960s and 1970s that didn't really seem to matter.

Today's computer environment is very different. With technologies and market conditions changing more quickly than ever before,
flexibility and rapid response are becoming increasingly important.
Companies must be able to change strategies and products quickly
and efficiently. In this environment, bureaucratic rigidity can be
deadly.

Former IBM managers tell horror stories about the IBM bureaucracy. "What people don't understand is that the process of getting an
announcement from IBM is torturous and complex," says Amdahl's
Joe Zemke, a former corporate director of marketing at IBM. "There
are literally tens of groups that can knock it out." Zemke explains that
a product working its way through the channels at IBM can get hung
up, or even shelved, for a wide variety of reasons: "from technology
to service, from manufacturing to pricing, to the fact that the U.S.
likes it but Europe doesn't."

IBM's bureaucracy can be particularly frustrating to IBM researchers. IBM has one of the strongest research organizations in the
entire world, including a flock of Nobel laureates. So why does IBM
introduce so few innovative products? One major problem is that the
most innovative IBM products get trapped in the laboratory: they

never see the light of day. A former IBM research manager once told me that only one in twenty projects at IBM research and development (R&D) labs ever makes its way out of the laboratory. IBM's bureaucracy squelches many of the researchers' best ideas before they ever get to the market.

MIPS president Bob Miller, who once directed IBM's research labs in Boulder, Colorado, believes there are two sides of IBM, giving Big Blue a "schizoid" personality. On one side is IBM's research and development effort. "That side tries to stimulate innovation and produce patents and publish papers. They see themselves as having a role in the technical community, to increase the knowledge base of data processing and information processing. All you have to do is look at the *IBM Journal*. Many of the ideas and new technologies at startup companies are derivatives of work that appeared in IBM publications." The other side of IBM, says Miller, is "the side that has to produce the revenue and the profit, the marketing and sales side."

Of course, all technology-based companies have these two sides. But at IBM, the split is sharper, more divisive. The bureaucracy is so thick that it becomes unbearable for many innovation-minded researchers. Time and again, they watch as their best ideas are watered down or thrown out entirely by the bureaucracy. "That's why I left, you see," says Gene Amdahl, once IBM's top computer designer. "I didn't leave because there was a golden opportunity elsewhere; I left because I felt there wasn't any inside."

Amdahl recalls the frustration of trying to develop new products at IBM: "I was asked to participate in several new products—new market task forces—in which they'd put about five or six of us together. Before I went in to the first meeting, one of the executives would take me to the side. This happened every time such a meeting occurred, and I was always given the same message. I was told: 'First, when you look for a new product area, you must bear in mind that it's extremely difficult to develop a new market. Second, it is extremely difficult to develop a totally new product. And third, when you try to do both, it's virtually impossible.' So you went in and looked for an emerging market that looked big enough to be approached with IBM techniques and where you could identify what the size of the market was, what the market needed, and what products would fulfill those needs."

This system made it very difficult to move forward with exciting new technologies. "I really left, I'd say, for ego reasons," says Amdahl. "I couldn't be proud of what I did when they were filtering out the technology at a rate where I knew we weren't doing anything that compared with what was possible. And I was sure that there were people on the outside who were going to be doing what was possible, such as Cray or Control Data, at that time."

Miller left IBM for similar reasons. He explains: "My major departure point came with IBM when I had a dream of what office automation ought to be, I mean literally a dream. I said, 'This is what it could be and this is how the IBM company ought to be able to put this together and make it happen.' " Miller took his ideas to IBM's top management, but all he got was frustration. One chief obstacle: his office automation plan would have required a coordinated effort among many different IBM divisions. "The top management of the company, which is just about evenly divided between sales guys and financial guys, could not see their way clear to how you could convert this huge bureaucratic organization to doing something that spans so many divisions. They were just impossible to deal with."

So Miller decided to leave IBM. John Opel, then president of IBM, tried to convince Miller to stay. He offered to make Miller vice-president in charge of office automation. Miller responded: "John, I've got thirty years until I retire. Making me vice-president of office automation with no direct line responsibility, I'll just be a very frustrated guy." Opel said that Miller could call him directly if he ever became frustrated. But Miller knew that wouldn't work. He told Opel: "You'll get tired of hearing from me."

Instead, Miller took his ideas to Data General, a much smaller company. Using the company's existing office automation products as a base, he developed a new system called CEO. "We were very successful and we competed and we gained a lot of market share on IBM," says Miller. "In fact, if you look at integrated office, we had a bigger share of integrated office than IBM had. And they couldn't respond. They could not respond. They couldn't go across ten different divisions, throw it all together, and make it happen."

In many respects, IBM's size and mind-boggling bureaucracy hold its most creative employees as prisoners. Amdahl and Miller were both driven out by the creative limitations they felt at IBM. And according to Miller, the bureaucracy lives on at IBM. "I interviewed an IBM guy not too long ago for a job and he was explaining his frustration," he says. "He told me that it took more than 250 signatures to get a product announced at IBM. And my guess is that 50 percent of the people who were signing had to be educated about what they were signing. I don't think it takes 250 signatures to get a product out at Tandem. And even though DEC is a little more bureaucratic than Tandem, I don't think it's close to 250 signatures."

IBM's product development process is slowed down further by infighting and rivalries within the bureaucracy. In many cases, IBM designers vie against each other for project approval. "At IBM," says Compaq's Rod Canion, "you've got many different groups with many different territories to defend, requirements to get accomplished, fighting with each other to see whose product wins out. And there's really no one group that's got enough visibility and understanding to solve the whole problem for them. So when they go through their process of selecting a product, they're unable to combine everybody's good ideas. Because in the IBM system, it's important that *your* idea wins, not somebody else's whose is a little bit better. It almost guarantees that IBM's going to move slower: there's more bureaucracy, more people, more different points to touch. They won't be able to come up with an ideal solution in any particular situation."

Floyd Kvamme viewed IBM's bureaucracy from the viewpoint of a salesperson. For years, he tried to sell semiconductor products to IBM. "You could never get advanced products designed at IBM," says Kvamme. "They wouldn't allow you to call on the computer designers. They didn't want you to find them. That's why the technology in IBM's designs was always a generation back." According to Kvamme, IBM's semiconductor operation in Fishkill, New York, made the decisions whether to make or buy particular chips. The computer designers had no say; they were out of the loop. Explains Kvamme: "Essentially, Fishkill would say to the IBM designer, 'These are the things that are available for you to design.' "

Challenger companies have begun to capitalize on IBM's slow response time and organizational rigidity. Newer corporations, born in an era of constant change, are better equipped to survive in today's environment. "IBM's very size and strength is the source of their greatest weakness," says Compaq vice-president Mike Swavely.

"That is where Compaq's strengths play in, where we can take advantage of the weaknesses. The bottom line is that IBM's development cycles are much longer than Compaq's because of the size of their company, and because of the complexity of their existing product line. As a competitor, I am sure going to take advantage of those weaknesses."

Swavely notes that IBM has many advantages in its competition with Compaq. "IBM out-resources us tremendously and out-market-impacts us tremendously," he says. Nonetheless, Compaq is holding its own against Big Blue. Swavely boils the reason down to a single word: *flexibility*. "In a rapidly changing marketplace," says Swavely, "flexibility is a better tool to have on your side than financial resources."

❑❑ Problem #2: IBM's InstalledTechnology

IBM's flexibility is limited not only by its bureaucracy but also by the technology of its installed computers. Whereas competitors can make use of the most modern technology available to create new computer architectures, IBM must keep one eye on the past, making sure not to abandon its loyal long-term customers. "IBM is constrained by the continuity of its own base," explains industry analyst Jonathan Seybold. "IBM can't do drastic things from a product standpoint because the whole ethic, the whole bond the company has with its customers is: 'We will not strand you. We will give you an evolutionary product.' That constrains what IBM can do."

The accelerating pace of technological change, symbolized by the microprocessor, has accentuated the installed-base problem for IBM. Innovations are now limited less by technology than by the imaginations of their creators. New businesses are flocking into the market, offering consumers new solutions to their data-processing needs. This presents a two-fold problem for IBM. Not only are there more competitors in the computer market, but the competitors are less constrained in their design choices.

Startup companies are not captive to old design decisions. They can use new technologies in the most innovative ways possible. They can create entirely new computer architectures and applications without worrying about compatibility with previous generations of machines.

IBM, on the other hand, is a prisoner of history. Big Blue still gets a huge portion of its revenues and profits from the 370 series of mainframe computers, developed in the 1960s. Clearly, IBM cannot abandon these customers. So IBM is caught in a trap. IBM's existing customers want faster, more powerful machines—but they also want the new machines to be compatible with the IBM machines they already own. They want to be able to continue using the software they have developed over the past twenty years, and they want to use the same peripherals. In short, they want to protect their past investment while also moving quickly into the future.

IBM is also caught with a mix of machines that is ill suited to today's environment. While most computer vendors are focusing on networks of small computers, IBM development and sales effort is still oriented toward large, stand-alone mainframe computers. According to one recent estimate, mainframes account for about 45 percent of IBM's sales and 65 percent of its profits. The mainframe market is now growing much more slowly than the rest of the computer industry. Clearly, the action is elsewhere. IBM is trying to adjust, moving toward networks, smaller systems, services, and software. But change is slow.

So Big Blue is in a bind. As it develops new generations of machines, it must balance the pressures of innovation and compatibility. How much of the old should be incorporated into the new? This type of thinking necessarily slows down the development process, and it leads to machines that are always trailing the leading edge of technology. Maintaining compatibility with the past involves sacrifices. Machines linked to the past are never quite as good or as fast as they could be. Indeed, IBM's actions tend to be defensive, concentrated on protecting established turf rather than forging new territory. "One of the great opportunities of competing against IBM is to out-innovate them by not being encumbered with standards or architectures which are no longer applicable," says Dave Martin of National Advanced Systems.

Thus, IBM ends up providing customers with a slowly evolving product line, while other companies leap generations ahead, providing systems designed for every specific need along the spectrum of computer applications. IBM's product continuity, one of its most reassuring characteristics for customers, is also one of its greatest weaknesses in the new technological era of the microprocessor.

❏❏ Problem #3: Incompatible Products

Today's computer customers want their computers to talk with one another. They want to be able to send data from one computer to another, and they want all of their computers to be able to share common databases of information. Indeed, as I discussed in Chapter 2, the ability to network computers together is a key to success in today's computer market. The success of Digital and Sun in recent years is due largely to the strong networking capabilities of Digital and Sun computers. Each company relies on a uniform, consistent architecture for all its computers, making it easy for customers to link different machines together into networks.

Networking, however, is an Achilles heel for IBM. Big Blue has many different families of computers, each with its own distinctive architecture. There is the 370 mainframe family, and the System 36 midsize family, and the PC personal computer family, to name a few. According to one count, IBM has nine different architectures among its computers. It's fairly difficult to link together IBM computers from within the same family, and trying to connect computers from different IBM families is very difficult. Each family has its own operating system and protocols, and none can understand the others.

"What happened was that each IBM group went off separately," explains former IBMer Bob Miller. "You have System 38s and System 36s and you have 4300s and 8100s. There is absolutely complete overlap. And not only that, but they don't talk. The systems don't have any capability of dealing with one another. You can't take a program from an 8100 and run it on a System 36. At DEC, the story is 'take any one of our machines and you can run the software all the way up and down the line.' At IBM, you have Series 1, System 36, System 38, 8100, 4300—all different."

IBM has tried to improve connectivity among its computers, but networking disparate systems is a tough job. And even when IBM succeeds, it is at great cost. Money and time that could have been spent on new product development are spent instead on figuring out how to link aging incompatible systems. "They always end up compromising," says Compaq's Swavely. "I've done product marketing for years, and I know what tradeoffs you have to try to make. You just cannot optimize when you have that kind of environment to deal with."

Why has IBM ended up with such a scattered collection of incompatible computer families? In some cases, IBM has tried to prevent its new computers from cannibalizing its older ones. If IBM makes a new generation of computers compatible with existing models, customers might abandon the existing models for the new ones. By creating new architectures, IBM segments the market, locking some customers into one family of machines, while capturing new customers with the new family of machines. Rather than adapt to the market, IBM has always tried to control the market.

Some industry observers believe that IBM's problem with incompatible systems arose partly from its worries about government antitrust actions. In 1969, the U.S. government brought an antitrust suit against IBM, accusing Big Blue of monopoly status in the computer industry. In anticipation of a government move to break up the company, IBM management reorganized the company into separate divisions offering independent product lines.

Former IBM manager Zemke believes that IBM's General Systems Division, headquartered in Atlanta, Georgia, was formed specifically in reaction to the antitrust suit. "It gave them a whole [stand-alone] organization, a whole thing to pass off," says Zemke. "They could have said, 'Gee, here it is, it's in the minicomputer business, it's in the low end, it's in the fastest growing part of the business, we'll let you split that off.' "

IBM's reorganization not only led to computers that couldn't communicate with one another, it led to *divisions* that couldn't communicate with one another. For several years Zemke was responsible for monitoring the interactions between the old IBM sales force and the new General Systems Division sales force. The project was called marketing cooperation, but the two groups were hardly cooperative. Zemke recalls: "When you set up two sales organizations and ask them to cooperate, it's like putting a prime T-bone out there and saying, 'You two guys share.' I would take this view to [then-chairman Frank] Cary and he would smile. I'd say, 'Frank, it's not working,' and he'd smile. I am convinced that he did it in anticipation of the antitrust action."

By the time the federal government dropped the antitrust suit in 1982, the damage was already done. IBM was left with a perplexing morass of independent divisions and incompatible systems. According to several competitors, IBM has been partially paralyzed as it tries to draw its organization back together.

"I've heard some people say that IBM really believed that they were going to be busted up with the antitrust action," says Apple's John Sculley. "They would never talk about it, but they wanted to at least be in a form where the company, if it were broken up, had some discrete and very powerful entities from a shareholder's standpoint. And then I've heard that IBM was as surprised as anyone was when the Justice Department dropped the case." Sculley notes a certain irony in the timing. Soon after IBM began breaking apart its organization in response to the antitrust suit, Digital began to demonstrate the importance of a common architecture throughout its product line.

❏ ❏ Problem #4: The Arrogance of Success

Companies, like people, can develop a certain mindset, a certain way of viewing the world. I like to say that it is not bigness that destroys companies but, rather, the "bigness mentality." Success and size have a way of translating into a culture of arrogance that blinds a company to the real needs of the customers and to the rising challenges of competitors. Countries can fall into the same trap. In his book *The Discoverers,* Daniel Boorstin points out that for centuries China felt no need to look beyond its borders for goods or knowledge, preferring to believe that it contained all answers within its walls. In an interview with *U.S. News and World Report,* Boorstin said that the theme of his book was that "the great obstacle to progress is not ignorance but the illusion of knowledge."

IBM's mindset took shape in the company's formative years, as Big Blue emerged as the dominant force in the new computer industry. Unfortunately for IBM, the computer industry has changed dramatically since those days, leaving IBM with a mind set that is ill suited to today's rapidly changing market.

IBM's mindset influences its approach to competition. IBM sees itself locked in a market-share battle with its competitors. Its goal is to win sales away from the competition. That approach was successful in the 1950s and 1960s, but the market-share mentality is no longer appropriate for today's newly emerging markets. With new technologies popping up every day, today's companies must focus on creating new markets, not sharing old ones. As Big Blue continues to cling to the concept of winning market share, startups are creating entirely new markets for computing. Today, hundreds of other computer companies, both new and old, are creating their own opportunities and building something new.

IBM has never learned to see markets as phenomena of the future rather than records of history. Market-share thinking erodes competitive positioning: the resulting strategies are based on what the competition is doing rather than on what products could be built in the future. Although improvements in market share look good on paper and make management feel secure, nothing is secure in the face of advancing computer technology. Technology presses forward and changes the definition and content of the markets themselves, thus undermining the old notion that market ownership equals industry domination.

The IBM mindset is also reflected in the company's arrogance, its feeling that it can do no wrong. Change is not easy for large organizations to accept. Faced with change, large organizations often become defensive, acting on the basis of historical models rather than on the way things *could* be. Leading companies in the U.S. steel, auto, machine tool, and semiconductor memory industries have all become arrogant and resistant to change. So has IBM.

IBM's arrogance is, perhaps, best illustrated by its pricing strategies. In many cases, IBM acts as if customers will be willing— *should* be willing—to pay more for the IBM brand name. Though that might have been true in the past, it is less true every day. Floyd Kvamme, venture capitalist and former president of National Advanced Systems, believes that IBM opened the door for competition by keeping its prices unreasonably high for its large systems. "Okay, why did all this happen?" Kvamme asks. "Why could STC [Storage Technology] take a major portion of the tape drive business from IBM? Why could a National Advanced Systems or an Amdahl exist? You've got to remember that IBM's price umbrella was unbelievable."

IBM felt it could get away with overinflated prices because it believed it *owned* its customers. IBM never priced to cost; it always priced to value. And when you believe you own your customers, you have an inflated perception of your value. Quite simply, IBM began to lose track of the value of its product offerings. Kvamme notes that one IBM machine, the 158, sold for $2.1 million in the late 1970s. But IBM's fully loaded cost of manufacture, according to Kvamme, was just $95,000. "Certainly you aren't pricing to cost when you're building something for a hundred grand and selling it for $2 million," says Kvamme.

This gap provided an opportunity for challengers like National Advanced Systems. NAS's first IBM-compatible machine competed directly against the IBM 158. According to Kvamme, NAS figured it could build its machine for $150,000, significantly more than IBM was spending. But NAS sold the machine for $1.5 million, about 30 percent lower than IBM, and still had plenty of margin for profit. "That's why our version was so successful," says Kvamme. "In the first year of NAS's shipment history, we went from something like $1 million in revenues to $100 million, and we made $40 million operating margin." As competition entered the market, customers began to realize that IBM was charging them an arm and a leg for its products. IBM lost many systems sales to its lower-priced competitors.

❏❏ Case Study: IBM Personal Computers

The story of IBM's personal computers serves as a good case study of the problems IBM faces today. By breaking many of its own rules, IBM quickly moved to the top of the personal computer field. But ultimately, the old IBM ways won out. Although IBM is still a powerful player in the personal computer market, it holds nothing near the dominant position that industry analysts were predicting just a few years ago.

At the start, IBM wasn't even sure whether it wanted to make its own personal computers. Bob Miller, who was at IBM at the time, says that IBM seriously considered buying a company that could produce personal computers at a lower cost than IBM could on its own. Many IBM managers felt that personal computers were "below our level." Miller explains: "The rationale made a lot of sense: we're a big system company. Every time we've tried to do small machines we've screwed them up." So one group that was exploring make versus buy options recommended that IBM buy its way into the personal-computer market. The group's top recommendation was for IBM to acquire Atari, the video-game company that was, at the time, moving into personal computers. After hearing the group's presentation, Frank Cary, IBM chairman at the time, asked a few simple questions:

"Is Atari the best?"

"No."

"Who is the best?"

"Apple."

"Why would IBM ever want anything but the best?"

That simple question rattled the presenter. Miller describes the scene: "This poor guy had a hundred flip charts to justify an Atari decision, and Cary asked one simple question. I mean, he had financials, forecasts, the whole bit, and Cary asked the very simple question: 'Are they the best?' It was as if the guy had never thought about that question. Cary pursued the queston of who was the best. The answer was Apple, but Apple was not for sale, and IBM would never do a hostile takeover."

Knowing how far behind it was in the rapidly developing personal computer market, IBM boldly broke its own bureaucratic rules. The company created a new unit, based in Boca Raton, Florida, to develop the IBM personal computer. The company selected Don Estridge* to head up the effort. For starters, the company gave Estridge a building in Boca Raton and generous financing for the project. But most important, the company gave Estridge complete control of the project, unprecedented freedom to manage the project as he saw fit.

Estridge ran the personal-computer project more like a startup company than an IBM operation. He developed a radical plan for a computer containing no IBM proprietary parts. Everything but the name would come from elsewhere. Estridge recognized that IBM did not have time to build a competitive personal computer from scratch, so he drew on the innovations of other companies. The PC's memory chips, its printer, and its disk drives were all provided by outside companies. IBM contracted with Microsoft and Personal Software to adapt software packages for the PC. IBM also broke its own distribution rules, making PCs available to the general public through retail channels.

At first, Apple greeted IBM with open arms, hoping that the presence of IBM would add legitimacy to the fledgling industry. Apple even ran an ad welcoming IBM to the industry. But within a matter of months, Apple and other personal-computer makers began to regret the arrival of IBM. IBM's PC took the market by storm, pushing aside all competitors. In 1981, Apple had 21 percent of the worldwide personal computer market. By 1983, it had slipped to 13 percent. Meanwhile, Big Blue's market share jumped to a solid 17 percent in 1983. Apple's Steve Jobs even spoke of the merits of breaking up IBM.

---

*Don Estridge and his wife were killed in an airplane crash in 1985. Estridge had become one of the best known and most highly respected executives in the personal computer industry and one of IBM's finest industry ambassadors.

Welcome, IBM.

Seriously.

Welcome to the most exciting and important marketplace since the computer revolution began 35 years ago.

And congratulations on your first personal computer.

Putting real computer power in the hands of the individual is already improving the way people work, think, learn, communicate and spend their leisure hours.

Computer literacy is fast becoming as fundamental a skill as reading or writing.

When we invented the first personal computer system, we estimated that over 140,000,000 people worldwide could justify the purchase of one, if only they understood its benefits.

Next year alone, we project that well over 1,000,000 will come to that understanding. Over the next decade, the growth of the personal computer will continue in logarithmic leaps.

We look forward to responsible competition in the massive effort to distribute this American technology to the world. And we appreciate the magnitude of your commitment.

Because what we are doing is increasing social capital by enhancing individual productivity.

Welcome to the task. **apple**

Reprinted with permission, Apple Computer, Inc. © 1983.

The PC's success was far greater than even IBM had projected. IBM was not alone in its underestimation. Almost no one had completely understood the incredible potential of the personal computer. One market research firm estimated that the personal-computer market would reach $2 billion in sales by 1985. In fact, it reached $19 billion.

The phenomenal success of IBM's PC was largely a result of the entrepreneurial, innovative spirit that came alive in Boca Raton. The entrepreneurial resurgence was short-lived, however. Analyst Jonathan Seybold explains: "Boca was formed as an independent business unit. It was more successful than anybody at IBM anticipated. It was clear that because Boca was the desktop workstation company, that Boca was going to be the tail that wagged the rest of the company. Boca became too important to IBM, not only from a revenue standpoint, but more important, strategically, to let it run by itself. So IBM had to bring Boca back into the fold. Doing that cost them a lot of time. Because when you do that, the innovation stops, the momentum stops."

According to Bob Miller, bureaucracy always wins out at IBM. "That huge bureaucracy couldn't deal with Estridge being an independent," says Miller. "They don't even like individuals within IBM getting too much attention. So he was violating every rule. They're very Japanese that way. I don't think it's practical to remain independent at IBM for anything other than a very short period of time."

As IBM's personal-computer operations became less independent, innovation was the big loser. IBM did not follow up its initial innovation with continued improvements. The company rolled out product improvements only gradually—leaving a huge opening for other companies. Challenger companies, not IBM, became the trend-setters in graphics, communications, and applications for the IBM personal computer.

Perhaps most important, IBM was very slow in developing the hardware and software needed for the PC to communicate easily with other IBM computers. By connecting PCs to mainframes, IBM could have increased the value of both machines. IBM management recognized the PC's potential for increasing mainframe sales as early as 1983, but Big Blue did not provide the necessary software until four years later. The slow-moving culture of the Big Blue bureaucracy cost IBM dearly.

Had IBM moved quickly to make improvements on the PC, the clone-makers never would have had a chance to make inroads into the market. "Think about it," says Sun's Scott McNealy. "In the first eight years of that product, what did IBM do to it? They didn't invest anything other than manufacturing engineering. And then they didn't even do that very aggressively. They did no operating systems enhancements; they did no graphics enhancements; they did no networking enhancements; they did no CPU enhancements; they did no application software for it. They laid the biggest goose egg for a golden goose opportunity. They did nothing to that product, no engineering."

McNealy continues: "It's all sales and distribution and marketing and advertising—and Charlie Chaplin running around with a flower. And none of it had to do with engineering. They were too sales-oriented. They could have absolutely buried Apple. The reason Apple is there is because they have a small amount of engineering in their product." After introducing the PC, IBM could have enhanced the product in many ways. It could have added a windowing environment to it or added some AI [artificial intelligence] capabilities. But instead, says McNealy, "they just sat back and let things happen to them."

IBM's lack of sustained innovation on the PC was particularly damaging because the PC was so easy to copy. The PC was completely constructed from standard, off-the-shelf parts and used an operating system that was "industry standard," not IBM proprietary, so *anyone* could buy the parts, package them, and sell a clone of the IBM PC. IBM had little choice in developing the PC. It had to use industry-standard parts to get the PC to the market quickly. But there was a downside to the PC's open architecture: IBM lost control. "In the past, IBM had complete control of what was inside of the processor, complete control of all of the operating system software, and in many cases, complete control of all the application software," says Compaq's Rod Canion. "When they wanted to make a change that would throw a wrench in the works of their competitor, they had the ability to do it. They lost that ability when they went to the open architecture and to the industry-standard supporting environment."

So the IBM PC was both hugely successful and easy to copy. Not surprisingly, clone-makers entered the market in droves, offering compatible machines at cut-rate prices. IBM cut its own prices in response, but it let the clone-makers take the lead in adding new features. As a result, the clone-makers grabbed a huge share of the market.

Indeed, IBM's personal computer has become a vehicle for other companies to ride to success. Witness Zenith's success in the government market, Tandy's success in the small-business market, and Wyse's success in the OEM market. These companies have been more aggressive than IBM in making improvements to the PC and targeting new markets for the machine. Today, when we look at a pie chart of the personal-computer market, we see that "Other companies," not IBM, owns the largest share of the personal-computer marketplace. These other companies have succeeded largely because of Big Blue's lack of sustained innovation.

The success of "Other" in the PC market has had a profound effect on customer perceptions. In buying PCs, many customers bought non-IBM computers for the first time. And they liked what they got. "As long as that experience was limited to lawyers' offices and doctors' offices, it wasn't so bad for IBM," says Amdahl's Zemke. "But when you have Fortune 1000 customers getting the experience of being able to buy plug-compatible, it's a different story." Zemke believes the success of PC clones has dramatically expanded the market for the plug-compatible mainframes made by Amdahl and others. The PC clones introduced customers to IBM alternatives at both a low risk level and a low purchase point. After experiencing cost savings at the low end, many customers began thinking about IBM alternatives at the high end as well.

❑❑ Big Blues for Big Blue

IBM's string of recent setbacks has, at last, begun to erode Big Blue's once-impeccable image. Customers are beginning to see IBM as vulnerable. The success of Digital in the Fortune 1000 market, IBM's traditional stronghold, has been particularly damaging. The fact that a competitor has even a chance to beat IBM on its own corporate turf tarnishes Big Blue's image. That Digital systems are actually outselling IBM's machines is image-shattering. IBM's weaknesses are now plain for the world to see. An article in *Electronic Business* magazine put it this way:

> [Digital] is in the enviable position of being able to sell against IBM's greatest weaknesses: Big Blue's hidden costs, its mainframe-based centralization, its proliferation of hardware families running incompatible software, its traditionally high pricing, and its chronic underachievement in scientific markets.

IBM has tried to respond to the challenges of smaller, more innovative companies. But in case after case, IBM is falling short. In the technical workstation market, for example, IBM tried to regain ground with a machine called the PC RT. But when compared to competing workstations from Sun, Hewlett-Packard, and Apollo, the RT's price/performance ratio just didn't measure up. As a result, crucial third-party software developers withheld their support. IBM eventually improved the machine's floating-point performance, but the damage was already done.

The future is sure to bring more, not fewer, challenges for IBM. Almost everyone agrees that the computer market will be rocked by one radical change after another during the next twenty years, and IBM is not well situated to compete in such a turbulent environment. Constant, radical change doesn't play to IBM's strengths. As Apple's John Sculley points out, radical change is likely to wreak havoc "for a company with a no-layoff policy, a company that has hundreds of thousands of employees, a company that has cultures that are very much anchored to where the company has been."

IBM is at a definite structural disadvantage when compared to its challengers. Challenger companies are "far more intent upon figuring out what is it they are trying to build as opposed to what are they trying to protect," says Sculley. "We can only expect more change, we can only expect that we are going to have to be more innovative. I see those as strengths of a company like Apple. IBM has some wonderful strengths, but being the leader in innovative products in the marketplace is not one of them. Being highly flexible is not one of them. It is not that IBM is going to fail, it's not that IBM isn't an extremely well-managed company, it's not that their customers are not very satisfied with what they do, and it's not that IBM won't be very important. But the important observation is that there is plenty of opportunity for the rest of us."

Next, in Part Two, I explore those opportunities, explaining how challenger companies can best capitalize on IBM's deteriorating position.

New Strategies for Success

For success in this environment [frontier America], the special-
ized skills—of lawyer, doctor, financier, or engineer—had a
new unimprotance. The rewards went to the organizer, the
persuader, the discoverer of opportunities, the projector, the
risk-taker, and the man able to attach himself quickly and
profitably to some group until its promise was tested.

Daniel Boorstin, *The Americans: The National Experience*

□ 4

Leadership

■■ PEOPLE LIKE TO DEAL with leaders. They like the security and the status. For many years, that was a major IBM advantage. When people thought about computers, they thought of the industry leader: IBM.

But that is no longer the case. The computer industry is no longer a single industry; it is a collection of subindustries. And in most of the subindustries, IBM is not the leader. When people think of educational computers or desktop publishing, they think of Apple, not IBM. When they think about engineering workstations, they think of Sun, not IBM. When they think about on-line transaction processing, they think of Tandem, not IBM.

The list goes on and on. The leader in supercomputers is not IBM. The leader in parallel processing is not IBM. The leader in laptop computers is not IBM. The leader in minicomputers is not IBM. In each of these subindustries, a challenger has grabbed the lead. Today's customers recognize Apple and Cray and DEC and Compaq and Tandem as leaders. These companies developed new technologies and created new markets—and therefore became recognized as leaders.

Leadership is a very powerful marketing weapon. Once a company has established a position of leadership, it is difficult to dislodge. In the 1960s and 1970s, IBM reaped the rewards of leadership. Other companies tried to challenge IBM, but customers stuck with the leader.

With the fragmentation of the computer industry into multiple niche markets, the leadership game has shifted. Now it is IBM that is trying to dislodge other companies. Today, IBM is a follower, not a leader. Increasingly, the IBM challengers are setting the tone and direction of the computer industry. The desktop-publishing market flourished before IBM gave its seal of approval. Engineering workstations gained credibility on the basis of Sun's products, not IBM's.

Indeed, IBM's presence is no longer needed to establish the credibility of a new computer market. Now, it is IBM that must scramble to gain position in niche markets. IBM has become the pursuer, not the pursued.

This shift in industry leadership has just begun. Some customers still view IBM as a leader—but perceptions are shifting quickly. The opportunities for challenger companies are enormous. In this chapter, I discuss how challenger companies can establish—and maintain—positions of leadership. Leadership in the technology-based industries is based on a complex combination of factors, including:

- Differentiation
- Vision
- Commitment from the top
- Taking risks
- Focusing on your strengths
- Being good at everything

After discussing each of these leadership factors, I present case studies of several challenger companies that have been particularly successful in positioning themselves as leaders.

❑❑ Differentiation

Differentiation is probably the most important factor in building a leadership position. All successful challengers establish a unique market presence: they launch distinctive products that change price/performance curves or start new segments of the computer business.

New companies must offer significantly new ways of doing things—or customers will simply continue to rely on established products. The key is to create a new niche and then fill it with a first-rate product. Altos Computer Systems, for example, created a niche by building a multiuser system that fit into the gap between personal computers and minicomputers. Altos sold its systems to small and midsized companies that couldn't afford minicomputers, but needed more power than personal computers could offer.

Cray is another example. Cray's initial plan was to build the world's fastest computer, and it has maintained that position since its founding. In the early 1980s, Cray management thought about making a slower, less expensive machine, but the idea was eventually discarded. Instead, the company has bet its entire future on its leadership position in the supercomputer market. Cray chairman John Rollwagen defended the strategy in a 1985 article in *Fortune* magazine:

> *I tell our employees that we have no backup plans. We're not going into peripherals or services or "affordable" supercomputers. We're staying with real supercomputers and are committed to the top end of that market, which means being technically the best and remaining creative and iconoclastic.*

New niches continue to arise, seemingly out of thin air. As recently as 1985, few if any business plans contained information on the growth of the desktop-publishing market. But by 1987 that market had reached sales of $749 million. By the time a new trend is recognized by the major players, it's often too late to establish a leadership position. Therefore, companies must establish leadership by recognizing a new market early, then filling a need through innovation. The key questions: How soon can you get to market, and how soon can you offer a differentiated product? Initial success in a niche market is essential for future growth. By winning in a specific market, a company can qualify for entry into other markets.

Market analyst Jonathan Seybold sees "niche-manship" as the defining characteristic of the new IBM challengers. "These companies are taking on IBM in a very different fashion than the mainframe computer manufacturers that IBM crushed earlier on," he says. "The mainframe companies took on IBM on its own turf, head-to-head, and they lost. These companies are, consciously or unconsciously, trying not to do that. They're trying for differentiation, trying for niche-manship with more closely focused products, but not going head-to-head with IBM. Each one of the companies has gone up against IBM, but it has done so by seeking a real product niche and real product differentiation. I've always thought that the losing strategy is to go up against IBM without a differentiated strategy."

To establish and maintain leadership, differentiation must be in terms of new ideas and new applications, not simply lower price. Companies who live by price will die by price. Many makers of PC clones fit in this category. They are positioned merely in terms of price. Sooner or later, they will need to start making more money. They will want to expand by offering more complete solutions. But when that time comes, they will find it very difficult to change their positions. They will always be viewed in terms of their prices.

As the number of niche markets increases, so does the number of potential challengers to IBM. Frankly, I believe that niche markets will continue spawning new computer companies for decades to come. That is bad news for IBM. IBM's overall position will erode as new technology allows challengers to create products and solutions for ever-narrowing segments of business. This is not to say that IBM will go away. But just as Gulliver was overcome by the Lilliputians, IBM will be attacked by a multitude of small but tough challengers, each a leader in its own right.

"In competing with IBM, virtually anyone else has the opportunity, by their very size and narrower focus, to do a better job of gleaning a specific application area and filling it better," says Dave Martin, vice-president at National Advanced Systems. "Our most important strategic thrust is to do that in the engineering/scientific area. We don't think that IBM can afford that kind of focus in that particular area, and we are filling it better."

❑❑ Vision

To become a leader, a company must see beyond the day-to-day details. It must have a clear vision of where its products are going and where the company itself is going.

John Butler, vice-president of Honeywell Bull, tells the story of two men who were laying bricks. Someone walked up and asked the men what they were doing. The first man answered: "I'm laying bricks." The second man answered: "I'm building a cathedral." Leadership companies have the vision to see that they are building cathedrals.

Leaders must also develop ways to *communicate* their visions.

Many companies try to communicate their visions to the world through advertising campaigns. But that is the wrong approach. Companies should communicate their visions *internally* before trying to spread the visions *externally*.

If a company's vision is well understood by the company's employees, the vision will get out to the world in a natural way. If salespeople understand where the company is going, they will communicate the message to their customers. Often the company's vision gets out in indirect ways, through the enthusiasm and pride of the company's employees. Even telephone operators can communicate a sense of pride.

Apple has developed an extensive "culture training" program for its employees. Employees learn not only about Apple's products, but also about Apple's approach to the world, its spirit, its vision. After going through the program, many Apple employees turn into Apple evangelists, spreading the Apple gospel through the world. I once stood in a supermarket line behind a woman wearing an Apple T-shirt. She radiated pride at being an Apple employee. When the checkout clerk asked about the T-shirt, the woman gave an enthusiastic sales pitch about Apple.

❏❏ Commitment from the Top

In technology businesses, leading companies are often differentiated by strong, visible leaders—leaders like John Young of Hewlett-Packard and John Sculley of Apple. The integrity of the leader reflects the integrity of the company. Digital, Hewlett-Packard, Apple, Tandem, Intel, and Sun are but a few of the challenger companies that are positioned by strong management at the top.

Indeed, leadership *must* come from the top of the company, from the chief executive and other top managers. Commitment from the top is important for two reasons. First, the top managers set the tone for the rest of the employees. They communicate a style and a message that permeate the rest of the organization. Second, the outside world often evaluates companies by their top executives. Top executives become a sort of filter: outsiders attribute value to a company on the basis of its top managers. People will take a company seriously only if they view the top executives as strong leaders.

Top executives can show their commitment in many ways. When the president of Milliken Textiles wanted to improve quality at the company, he didn't just issue a directive to the employees. Instead, both he and the chairman of the company dedicated 50 percent of their time to quality-related projects.

Jimmy Treybig of Tandem is the epitome of a committed leader. Treybig is wealthy enough to retire for life, but he continues to work as hard as he did when he started Tandem. Although Tandem has grown into a $1 billion company, Treybig finds ways to stay in touch with Tandem's employees. He goes out of his way to meet new employees at the company, and he encourages employees to come to him with suggestions and complaints. Tandem was one of the first companies to establish its own corporate-wide television network for employee training and customer communications. In addition, everyone in the company is linked into an electronic mail system, so anyone can send electronic mail to Treybig. "If we had a bad manager in Des Moines, anyone there can send me a mail message, and no one else ever knows," says Treybig. "It all revolves around having people involved, having them care, having them know that the president is a human being and having ways to express their frustration."

Treybig recognizes that electronic mail is a limited form of communication. To address the need for direct interpersonal contact, Treybig set up a program called Tandem Outstanding People (TOP). The TOP program brings together a small group of Tandem employees for a type of retreat. Each group includes people from different functions, backgrounds, and locations within the company. "TOPS brings the company together," says Treybig. "The program allows us, in a social and business environment, to present our future strategy. It creates an internal network. It creates respect for teamwork, so that a salesperson knows a service person who knows an assembly worker who knows an accounting clerk who knows a manager. And they all know vice-presidents and me, because we're all involved in TOPS too."

Treybig continues: "I can't go to Des Moines very often, but I know the best people there. If I walk into that office, I know two TOP winners probably. I don't know them really well, but I drank beer with them or something, maybe partied with them, maybe danced with them."

□□ Taking Risks

To thrive in fast-changing markets, you have to take risks. There will always be someone out there who is taking risks, and the ones that succeed will drag customers away from the companies that don't take risks.

Leaders seem to relish risk. They like to push ideas to the surface and make them happen. Creating new things is what leadership is all about. Opportunities are not just sitting there waiting to happen; leadership companies have to take risks and make them happen.

Risks can take many different forms. Many risks, of course, are technological risks. Companies invest tens of millions of dollars in new computer architectures, not knowing whether the new architectures will really work, or whether they will meet the needs of the market. Other risks are nontechnological. When Unisys and Sperry decided to merge, joining two long-time competitors, that was a major risk. Similarly, Honeywell, Bull, and NEC took a risk when they decided to work together. But as always, the risk is balanced by a potentially big reward: if the Honeywell–Bull–NEC joint effort succeeds, it will have access to all three major computer markets (the United States, Japan, and Europe).

The ethic of risk-taking is one thing that distinguishes technology-based industries from other industries. John Sculley noticed the difference when he moved from Pepsi to Apple. "The risk of failure would be almost unthinkable and unspoken in corporate America, the world I came from," says Sculley. "No one ever spoke of failure." In Silicon Valley, the situation is very different. The risk of failure is always in the air. "We've seen a whole new set of values emerge," says Sculley. "We all believe passionately in the future, but it hasn't happened yet. We are willing to bet the entire company on those visions. But when we're wrong, there is always the chance that the company will lose its reason for existence. There isn't a company in the computer industry today that hasn't had to go through those kinds of issues in one form or another."

Although (or perhaps because) the risks in computer business are large, the industry tends to be more forgiving of failure. In most industries, leaders of failed companies are never heard from again. But in technology-based businesses, the situation is very different. "In Silicon Valley, if someone fails, we know that they're in all likelihood going to reappear in some other new company in a matter of months," says Apple's Sculley. "We've got an entirely different set of values in the information technology industry. People tend to be big risk-takers, but they expect bigger rewards in return. They are oriented toward change. In fact, they thrive on change and look at it as a way of uncovering new opportunities."

❏ ❏ Focusing on Your Strengths

To become leaders, companies must dedicate themselves to particular areas of expertise. Tandem, for instance, focuses on fault tolerance and transaction-processing applications. Sun focuses on open systems and UNIX-based environments.

Companies can't develop all technologies: there are simply too many different technologies in today's computer industry. Each of these technologies progresses at a different rate. If a company tries to develop everything by itself, it will be good at nothing. To become a leader, a company must acquire some technologies and develop only those technologies where it has industry-leading expertise. Apple, for instance, buys many of its subsystems and peripherals, but it develops its own systems software and graphics interface. In the 1970s Apple bought its disk drives but created a breakthrough drive controller that pushed performance of the Apple II far ahead of the competition.

Too many companies have a "not invented here" mentality. They try to do everything themselves—and, as a result, don't do anything all that well. A company that spreads itself too thin will constantly reinvent the wheel and will never develop a leadership position in any particular area.

□ □ Being Good at Everything

While companies should focus their development efforts on only a few technology areas, they must be good at *all* aspects of running a business. A company with great product development but poor marketing will fail. Similarly, a company with great marketing but poor product development will fail.

The marketplace is full of piranhas: they will attack at the slightest sign of weakness. If the quality of a product slips, it will be displaced by another (perhaps technologically inferior) product. But if the new leader slips in its marketing or service or distribution, it too will be displaced. The only way to maintain leadership is to be good at everything.

□ □ Case Studies

Increasingly, the challenger companies are grabbing the leadership positions in the computer industry. They are setting the directions, opening new markets. IBM is being forced into formulating strategies for competing with *them,* not vice versa. IBM is in the new and uncomfortable position of saying, "We've got to compete with Tandem in transaction processing, so we'd better go make a deal with Stratus." It isn't challenger Tandem saying, "Oh my God, how do we reduce our price?" Tandem has continually moved forward. It has advanced the technology and developed the market. It has a very aggressive strategy. IBM must react just to catch up.

Tandem, Compaq, Apple, and Sun are the new models for computer companies. There was a time not too long ago when every computer company wanted to be like IBM. No more. Everyone believes it can do things better.

Next, I present brief case studies for several companies that have emerged as leaders in the new computer industry. Each company has used a different strategy, but each has grabbed a leadership position in some slice of the computer industry.

❏ *Sun Microsystems*

The founders of Sun Microsystems took on a difficult challenge: they wanted to build a general-purpose computer, in an industry already crowded with general-purpose computers. "By definition," says Sun president Scott McNealy, "that puts us in competition with some part of DEC, IBM, HP, Data General, Wang, Prime, and even some of the personal-computer companies. We aren't a specialized computer maker like Computervision or Daisy. We're not in a vertical market, competing based on applications software. We're competing on systems software and hardware architecture and that sort of thing. So from that perspective, we had a significant challenge: the world didn't need another computer company; there were plenty of them out there."

To differentiate Sun in an already crowded market, McNealy and his colleagues decided on a new type of computing architecture. McNealy notes that there were previously two dominant architectures. First was the mainframe-based, hierarchical architecture championed by IBM. "That's not a fun game. You have to go compete under IBM's ground rules," says McNealy. The second approach was the minicomputer-based approach, pioneered by DEC, HP, Prime, Data General, and Wang. In that approach, many users share a minicomputer on a time-sharing basis.

Sun pioneered a third approach, based on networks and microprocessor-based workstations. The Sun approach extends the idea of distributed computing—distributing more computing power to individual users. In this approach, each user has a powerful personal workstation. These workstations (along with shared "server" computers) are all linked together via a network.

"In IBM's world, the mainframe is the hub," explains McNealy. "In DEC's world, the minicomputer has been the hub—although they're moving much more aggressively to distributive computing than IBM is. And in Sun's world, the network is the hub."

The network-based approach can provide important performance advantages, since each user has a powerful workstation on his or her own desk. But users can still get economies of scale by sharing expensive peripherals like printers, disk drives, tape drives, simulation engines, and other output devices. Says McNealy: "That gives us a price/performance advantage over the mainframe model."

Sun also had a second element in its differentiation strategy. Sun's improved price/performance provided users with an immediate benefit in choosing Sun systems. But Sun also wanted to position its products as a "safe buy." To do that, the company committed itself to using standard operating systems and protocols. As I discuss in Chapter 9, conforming to industry standards is becoming increasingly important in the computer industry. Since Sun's machines conform to industry standards, customers know that their equipment will always be compatible with products from other vendors, even if Sun should fail in the long run. In addition, Sun strives to make industry standards itself by licensing virtually all of the technology it creates. By making its technology widely available, like its Network File System (NSF), Sun has created de facto industry standards and raised its own credibility in the marketplace.

This two-pronged strategy has been an incredible success. Sun has established itself as a powerful force in the desktop workstation market. Six years after its founding, Sun has reached $1 billion in annual revenues and is fast becoming the next major computer company.

❏ Hewlett-Packard

Around 1980, Hewlett-Packard president John Young took a hard look at HP's computer business, and he didn't like what he saw. Although HP had been in the computer business for more than a decade, it was still somewhat of a specialty supplier. It did not offer a full range of products to meet customers' needs. More troubling, HP did not seem to have the right technology, expertise, or corporate structure to become a leading full-range computer manufacturer. The company's small, entrepreneurial divisions, which had helped make the company such a success in the scientific-instrument business, were not well suited to the computer business. Customers wanted their computers to work together, so a unified company-wide approach was needed.

Young realized that he had to make some changes. Young himself, like many top HP executives, had risen through the instruments side of HP. No one at the top of HP had any real systems-level expertise in computers. And HP's R&D labs still focused primarily on test-and-measurement technology, not on systems-based computer technology.

So Young made a bold move: he reached outside the company and recruited Joel Birnbaum to take over as HP's director of computer research. Birnbaum had directed computer science research at IBM's Watson Research Center for five years, and he had a reputation as a maverick. He didn't quite fit into the HP mold, but HP needed some new blood and new ideas.

Birnbaum steered HP on a radical course. He argued that HP should work toward a single computer architecture to unify all of the company's computer products. That meant some big changes in HP's organization and mindset. Engineers working on alternative computer architectures had to shift direction. And division managers, who previously had had near-total autonomy at HP, now had to begin coordinating their activities with managers in other divisions.

In choosing a uniform architecture, Birnbaum worked within a few constraints. He aimed for:

■ Price/performance leadership
■ Scalability across a wide range of implementations
■ Low cost of ownership
■ Straightforward migration from the current HP line

Working within these constraints, Birnbaum and his team decided to focus on RISC (reduced instruction set computing) technology. The RISC design philosophy leads to processors that are much simpler and faster than conventional computers. RISC minimizes the number of primitive operations that a processor can perform, eliminating the complex special-purpose commands that bog down traditional processors.

As part of its move toward uniformity, HP also decided to phase out its proprietary networking schemes and to base all its networking products on an emerging industry standard known as OSI (Open System Interconnect). With OSI networking, HP machines could communicate more easily with machines from other manufacturers. HP's early support of OSI consolidated its position as a leader in promoting industry standards.

To back up its new technological directions, HP realigned its organization. Previously, HP had been organized along product lines. But that organization did not mesh with the company's new uniform approach to computer systems. So HP centralized development for all product lines, while assigning divisions the marketing responsibilities for particular groups of customers.

With its new technology and new organization, HP has emerged as one of the hottest computer companies of the late 1980s. And it has proved that you do not have to be a startup company to become an innovative leader in the new computer industry.

❑ Cullinet

The case of Culprit, a corporate-report generator from Cullinet, clearly illustrates the importance of product differentiation in achieving leadership. Culprit started out as a weak product. But with some careful repositioning—including a name change—it turned into a big winner.

In 1971 Cullinet introduced its first product, Culprit, a software package running on IBM mainframes. The company had just raised $500,000 on Wall Street, and its future hinged on the success of Culprit. For a while, that future looked pretty bleak. "We had run out of money," recalls Cullinet president John Cullinane. "We had a payroll due one day of $8,500, and we received a check that day for $8,500. We were a one-product company. We were running out of time, we had no technical resources, no time, and no money left. What do you do?"

With the prospect of failure closing in, Cullinane came up with an idea: Why not target a version of Culprit specifically at EDP (electronic-data processing) auditors? Some auditors were already using Culprit, and the product seemed well suited to auditing tasks. So Cullinane developed a new version of Culprit called EDP

Auditor—which was really little more than a renaming of the original product. EDP Auditor could be used for statistical sampling and audit verification notices, as well as for training and support geared specifically to auditors. Although the product differed little from the original, Cullinane sold EDP Auditor for about $20,000, roughly twice the price of the original Culprit version.

At first, some of Cullinane's colleagues were skeptical about EDP Auditor. But Cullinane pushed ahead with his repositioning strategy, creating a users' group for EDP Auditor. The users' group became a clearinghouse for new ideas: users shared information and devised more applications for the system. EDP Auditor became a big hit within the auditing community. The product gave auditors something they were looking for—independence of data-processing departments. That independence meant a greater sense of integrity for the audit.

One interesting fallout of the success of EDP Auditor was that the original Culprit version started selling well too. "Auditors used our product so remarkably well that they embarrassed the data-processing department," explains Cullinane. "The question coming from the management was: 'How come the auditing department can do in three days what you guys say takes three months to do?' So then the data-processing department would buy the Culprit version!"

Through repositioning and renaming its product, Cullinet was able to sell many more report generators at $20,000 than it could at $10,000. The difference was that the $20,000 version was aimed at the right audience. Says Cullinane: "There were a lot of report generators at the time. This was a case of positioning and creating a differential: that's the key. The data-processing people couldn't make up their minds about which report generator to buy, but the auditing departments wanted our version. So by addressing auditing, we were then able to break through and sell a lot of EDP Auditors at $20,000 and make money."

According to Cullinane, Cullinet's history is filled with similar stories. "I've learned through the years that to survive and prosper, you have to roll out a new strategy at least every two years," he says. "And that strategy ought to include sufficient new products to allow you to meet your sales goals, and it ought to be presented in such a way as to put your competition on the defensive. I always feel that if I'm in a sales situation and I'm having to answer questions, that means that I'm being put on the defensive by my competitor and he's doing a better job than I am. The issue is how to position your products vis-à-vis the competition."

Cullinane continues: "If you were to look into Cullinet's history, you'd see that every two years, almost literally, Cullinet rolled out a new strategy that had those characteristics. I could go back to 1972, all the way up to 1983, every two years would have a different position or a different emphasis."

❑ *Amdahl*

The case of Amdahl is more of an exception than a rule. Amdahl has succeeded not by finding a niche, but by going head-to-head with IBM in the mainframe market, IBM's traditional stronghold. Amdahl's computers are designed to support the same software and peripherals as IBM mainframes: anything you can plug into an IBM mainframe, you can plug into an Amdahl mainframe just as well. In fact, Amdahl pioneered the idea of "look-alike" or "plug-compatible" machines, long before companies like Compaq began making clones of IBM personal computers.

"We probably compete more directly with IBM than any other company in the world," says Amdahl president Joe Zemke. "IBM's most powerful base is the Fortune 1000 customers, the big CPU [mainframe] users. IBM has been working with them the longest, and they have the biggest software investment and the most physical presence on the site. Most of the folks that are running those companies grew up with IBM."

In this competitive environment, Amdahl has differentiated itself primarily on the basis of performance. Indeed, the first Amdahl computer was an exceptional product. "It was four times more reliable than IBM, four times quicker to fix, and it installed in 20 percent of the time. More important, it was 60 percent more powerful than the biggest IBM machine available at the time," says Zemke. "Somebody once said if the IBM sales force had had the Amdahl 470, they would have wiped us out. It was that much better. We needed that to get the business started." Amdahl has continued with much the same strategy: its machines have stayed one step ahead of IBM in performance.

Zemke, who spent eighteen years with IBM before coming to Amdahl, believes that head-on competition with IBM actually has its advantages. "I've concluded that you're better off competing with IBM when you know where they are," he explains. "I've watched little startups that had a niche that IBM wasn't in. They work for two or three years, build their technology, build market awareness, get started, and get themselves to where the payoffs are. And then I watch IBM. Just a rumor that IBM is going to come into the marketplace knocks the startups on their ears. So I've concluded that maybe you're better off going head-on with IBM. So that's what we do [at Amdahl]—and we're never confused about who the competition is!"

This strategy has worked for Amdahl, but it is a risky one. Challenger companies will find many more opportunities differentiating by market or architecture or application, rather than differentiating on performance alone. As Sun's McNealy says: "I don't think anybody is going to go out and try to be a new Amdahl."

❑❑ Staying on Top

To a certain extent, leadership can be self-perpetuating. In general, people want to work at leadership companies. So leadership companies have the ability to attract the top technical talent. First-rate students from MIT and Stanford want to work at leadership companies. Technical employees have a lot of mobility and options today, and they want to go where the action is.

"In our industry, leadership is synonymous with excellence," says John Butler of Honeywell Bull. "Customers want to deal with leaders. The press wants to write about leaders. College kids want to work for leaders. And investors want to buy the stock in leaders."

So there is a type of positive feedback loop. Once a company emerges as a leader, it can attract top talent, which helps the company to sustain its leadership position. But leadership companies cannot afford to rest on their laurels. In today's fast-changing markets, companies can gain leadership quickly, but they can lose it quickly too.

Command large fields, but cultivate small ones.

Virgil

□ 5
Innovation

■ ■ THE PROVERBIAL MAN on the street probably views IBM as an innovative company. And until recently, many businesspeople probably agreed. But this image is simply not based on recent reality: IBM has been more of a follower than a leader in product innovation. IBM has never been very successful in developing truly innovative products. This fact is finally sinking in. In a 1987 poll conducted by *Fortune* magazine, corporate executives ranked IBM *below average* in innovativeness. They rated IBM 158th out of 306 companies in the survey.

The true innovators of the computer industry today are the challenger companies. Indeed, for every company I researched for this book, I found that sustained innovation was key to its success. The successful challengers didn't just invent things once, they sustained innovation over a period of time. If you really look at Digital, Cray, Apple, Compaq, Tandem, and the others, you will find a history of sustained innovation.

To be sure, everything else counts: marketing counts and people-oriented management counts. But it's innovation that creates the marketing power and drive. Without fundamental sustained innovation, none of these companies would have succeeded. Successful companies never let up. They're always working on improvements, always continuing to push. That's what keeps them ahead. These innovative companies are changing the way things are done technologically, while IBM lags behind, balking at changes that would flatten profit curves for its existing product lines.

Don Clifford and Dick Cavanagh highlight the importance of innovation in *The Winning Performance,* their book about high-growth companies. They write:

Innovation—the applied art of the new and better—underpins the strate-gies and successes of the winning performers. These companies innovate early and often—creating new markets, new products and services, and new ways of doing business. Almost by definition, innovation means breaking the rules and overturning conventional wisdom. It's the oppo-site of imitation and of business-as-usual. For the new competitor who seeks to survive and succeed in a market where established competitors have the advantage of scale, long-standing customer relationships, reputation, and financial staying power, innovation isn't just a nice-to-have. It's a necessity.

As I see it, there are two basic types of innovation: *radical change* and *incremental improvement*. Radical changes are very rare: the first automobile, the first instant camera, the first personal computer. These innovations are very important, of course. But we don't see many radical innovations in our lives; they come along perhaps once in a decade.

Most innovations are incremental improvements. But by calling them "incremental," I certainly do not mean that these innovations are insignificant. Incremental innovations can open huge markets. The automatic transmission in the automobile, for example, was an incremental improvement that extended driving to many, many more people. In the computer business, incremental innovations have dramatically broadened the audience and the applications for computers.

Incremental innovation is not simply a matter of adding a few new features. Incremental change is really valuable only when it draws new audiences to use technology. Too many technological innova-tions today are simply feature improvements. They do not extend the technology to new audiences and they do not create new markets. For example, the addition of sound capabilities to personal computers is a nice feature, but it offers no real utility.

Innovations can have incredible lasting power. Consider the Apple II computer. It drove the growth of Apple for nearly a decade, until the Macintosh took over as the company's keystone product. The 8086 chip played a similar role at Intel.

Innovative products, however, do not last by themselves. Innovation must not end with the initial product introduction; it must be a *sustained* effort. After a company brings a product to the market, it must continue to update and improve it. Technology products are never born perfect; they must be continually adapted to meet changing market requirements. All successful products go through a series of incremental improvements throughout their life cycles. When a company stops innovating, it stops succeeding.

Continuing to improve the functionality, production, and support of a product is the most difficult part of growing a business. Some companies, like Compaq and Sun, have done it extremely well. So has Apple. Apple's investment in the highly automated Macintosh Factory was an important incremental innovation: it paid huge dividends in the cost and quality of the product. Sustained innovation has also paid off for the Apple II family of computers. Third-party software companies and hardware vendors have continued to make numerous incremental improvements to the Apple II, adapting the computer to many different markets. Apple, meanwhile, made improvements that reduced manufacturing costs.

Successful innovations inevitably draw competition—which, in turn, forces the innovators to be even more innovative. When Convex created the minisupercomputer market in 1985, there were perhaps one or two other companies in the United States developing products for that market. But as soon as Convex's position as a technology leader was established, the company was immediately challenged to innovate again. Just one year after the introduction of Convex's C1 computer, more than twenty-five companies claimed to have products that fit the market.

As competitors begin nipping at their heels, technology leaders must keep innovating to retain their positions, their differentiation, their profitability. They become better in order to survive. Thus, single innovations are not enough. In order to succeed in today's market, sustained innovation, in the form of incremental improvements, is essential. And as product life cycles shrink, the pace of innovation must accelerate. Today, many technology-based companies must improve their products in a significant way every two years or so—or they risk losing their leadership positions.

This chapter examines the process and importance of innovation. First, I look at some models of successful innovation, presenting case studies from two challenger companies. Then, I examine some of the cornerstones for successful innovation—and describe what you can do to foster innovation at your company.

❑❑ Innovation as a Way of Life

Unlike IBM, challenger companies thrive on innovation. In many cases, challengers have created new markets with their innovations. The Cray I created the supercomputer market, the Digital PDP-8 created the minicomputer market, the Apple II created the personal computer market, the Convex C1 created the minisupercomputer market, Calma created the computer-aided design market, Mentor Graphics created the computer-aided engineering market, and Compaq created the IBM-compatible portable-PC market. According to International Data Corporation, the first four of these innovations created markets totaling $34.4 billion as of 1986, far in excess of the market created by IBM mainframes.

At challenger companies, innovation is a constant drive, a way of life. Apple chief executive John Sculley spends more time with engineers than with marketing people. Everyone at Apple is preoccupied with technology and innovation. Apple presented its vision of innovation in its 1984 annual report:

> *For Apple, our identity is innovation, and our vision is our belief that a computer is first and foremost a desktop appliance. To us, innovation means bringing computing power to people who have never used computers before. It means giving people a tool so they can create more easily, think more clearly, communicate more effectively in their jobs and in their lives. To us, it means giving people the freedom to realize their own potential through technology.*

Challenger companies are free of many of the constraints that limit IBM's innovation. Young companies, for instance, aren't bound by the constraints of multigenerational products. Unlike older, established companies, they don't need to worry about preserving compatibility with outdated product lines or with products from other divisions of the company. At IBM, it takes a great feat of cross-divisional coordination and diplomacy to make products compatible with one another. By contrast, young companies can focus on functionality and performance, not compatibility, when they design new product features. Thus, while change is often IBM's enemy, it is an opportunity for challenger companies—an opportunity to create advanced computing products.

In almost all cases, innovation is what determines success and failure at the challenger companies. According to John Cunningham, former president of Wang, sales at Wang grew rapidly on the strength of innovative products—then began to slip when the company lost its technological edge. "If you are able to innovate with the technology, you can grow a lot faster than the other guys," says Cunningham. "That is why Wang went from $100 million in 1976 to $2 billion seven or eight years later. It was delivering to the marketplace a leading-edge product with very good functionality, easy to use. And it put the bits and pieces behind it to get the thing delivered into the marketplace. If you fall off the strategy and the products no longer become leading-edge products so that you can deliver more functionality than the other guy, the strategy breaks apart." At the time, Wang was designing word processors created specifically to help secretaries be more efficient in their work.

Cunningham believes that all troubles at technology-based companies can be traced back to problems with technology and innovation. "Everyone who has failed in this business did it to themselves," he says. "They all had the opportunity to do the right thing. Some of them did and a lot didn't. Behind all of the current financial problems of many companies is a scenario, I believe, that goes like this: if you look at 1987, the companies are looked upon as bad financial performers, if you go back to '85–'86, the salespeople were a bunch of bums. If you go back to '83–'84, they were bad development decisions. Bad development decisions end up in weak orders, work through the backlog, no more revenues, expenses out of line, you lose money. But that is what happens. It is very simple. The companies who have lost their technology take three or four years to feel it on the bottom line. That's what happened. It's a very simple formula."

Innovation can take different forms at different companies. Next, I look at two of the most innovative challenger companies—Tandem and Cray—and describe how innovation drove the companies to success.

❏ *Tandem*

Innovation is at the core of the Tandem success story. Tandem's initial product, the Non-Stop computer, was unlike anything on the market—and it met a very real market need. Jim Treybig, Tandem's founder, recognized that people using computers for on-line transaction processing (such as airline reservations and stock transactions) had an enormous need for computer reliability. The Non-Stop computer met this need; it was designed with built-in redundancy so that it would never fail. The computer was an instant success with bankers, retailers, and others who needed fail-safe processing. During its first four years, Tandem doubled in size every year.

Tandem's initial innovation put the company in an enviable position. Because Tandem's first machine was so advanced, competitors found it very difficult to play catch-up. To shorten development time, a few companies tried to build competitive computers based on microprocessor technology. But they ran into problems with software development. Tandem's lead in software was so huge that no one could catch up in a short time. That created a financial barrier for competitors: to challenge Tandem, a competitor needed to start with enough money to fund a long software development effort. Tandem went largely unchallenged for many years.

While others struggled to catch up, Tandem did not stand still. It continued to make incremental improvements to its computers. Tandem spent more than $10 million developing a set of proprietary chips to replace the off-the-shelf chips used in its initial models. The new custom chips improved the performance of Tandem's machines, making them even more reliable and modular. At the same time, the chips made Tandem's technology more difficult for competitors to copy.

With its one-two punch of strong initial innovation plus sustained improvement, Tandem remains comfortably ahead of its competitors in the fail-safe computer market.

❏ *Cray*

At Cray Research, the maker of the world's fastest supercomputers, innovation means continually changing the rules of the game. As soon as other companies start to gain ground on Cray in supercomputer technology, Cray takes off in a new direction, bringing new approaches to supercomputer design. It's all part of an innovation-based strategy designed to keep Cray ahead of the competition.

"The longer we play with the same rules, the more they catch up with us," explains Cray chairman John Rollwagen. "They really polish their product and develop a lot of momentum, and they start coming in this direction. So we have to go off in another direction. Then they go shooting off—you know, like a cartoon picture—shooting off the end of the cliff. But we're not there, we're over here having a good time in this patch. So we have to live by our wits. We have to change the rules."

Cray has been an innovation leader ever since Seymour Cray built the world's first supercomputer in 1976. Cray's first computer was based on a dramatic new design. To increase the speed of the machine, Cray shortened the lengths of the computer's internal wires with a unique configuration scheme. These shortened wires generated more heat than in traditional computers, so Cray ran liquid freon throughout the machines to dissipate the heat. Later Cray machines increased speed even more, with still closer interconnections, and a different liquid fluorocarbon to cool the wires.

Cray has also been an innovator in his approach to the design process. Cray worries that previous designs might limit or constrain his thinking as he tries to move off in new directions. To avoid this problem, Cray actually destroys the plans for his previous creations before starting on a new project.

Cray's approach has been enormously successful. Cray's current supercomputers have the power of 50,000 personal computers working together simultaneously. With Cray's machines, scientists and engineers not only can do things faster, but can recreate events in three-dimensional form. For example, engineers can see, via precise computer simulation, the flow of air over a newly designed airplane wing. And auto manufacturers can "test" their new models by smashing computer-simulated cars into computer-simulated brick walls.

Cray is so innovation-rich that the company's biggest challenge is not encouraging innovation but figuring out a way to manage it. Cray is organized into small autonomous groups, with fewer than twenty-four people in each group. In all, there are about two hundred different groups. "The way we're structured, we're granular, we're project-oriented," says Rollwagen. Such an organization presents management challenges. "If you've got two hundred groups completely absorbed in their own projects, you have great potential for anarchy and diffusion and no focus," says Rollwagen.

As Rollwagen sees it, his primary responsibility is to bring a little order to the chaotic process of innovation. "My main job is to see that a vision is set for the company constantly, that it's examined constantly, and that it is preserved and then held high," says Rollwagen. "It's just like those Japanese tour groups, where they raise the flag and say: 'This is it!' Right? We all understand it. Then you all have an opportunity to decide whether to stay in this group or go to some other group, but this is why we've come together here."

Rollwagen asks researchers to monitor their own work and to reflect on whether it serves long-term company objectives. "If your answer is 'Yes' and you're happy with it, then I am happy too. If your answer is 'No,' that's not necessarily bad. But you've got to understand that at some point it's just not going to work. So it's your choice. You decide what you want to do. If you want to do something else, that's great—have a terrific time and write occasionally."

❏ ❏ Five Keys to Innovation

It would be nice to offer a simple recipe for innovation. Unfortunately, that goes against the very spirit of innovation, since true innovation requires breaking the rules and creating something entirely new. In short, there is no "correct" recipe, no magic formula. Nevertheless, there are some themes that emerged from my talks with successful challenger companies. The challenger companies have all created fertile environments for new ideas and innovation. How do they create such environments? Here are some general guidelines:

■ **Think small.** Big ideas are fine, but big development groups aren't. Many companies act as if throwing more people at a development project will guarantee its success. In fact, the opposite is more nearly true. The best and most innovative products generally come from small development groups. Why? Within small groups, it is easier to coordinate activities and keep the project on track. What's more, individuals working within small groups are more likely to feel a sense of commitment and a shared sense of purpose.

At Wang, for instance, a group of just five people developed a set of products that generated about $7 billion worth of sales for the company. "There were five people that did this product line—five people for the first twelve months, and they added about nine people to get the product up to the release date," says John Cunningham, Wang's former president. "Those were the products that basically drove the revenues behind Wang." Many people have blamed the IBM PC for seriously damaging Wang's business, but Cunningham has a different explanation. He notes that Wang lost all five of the key designers who had been driving innovation at the company.

MIPS president Bob Miller agrees that small, high-quality groups can make bigger impacts. "If you can grow your company and keep the resource number small, and the quality high, you've really done something," he says. "I think the more people you put in a company, the more trouble you're asking for in terms of your ability to respond."

Even more important than small groups is "small thinking." In most cases, bigness doesn't destroy companies: thinking big does. When companies think big, they tend to lump customers into broad statistical categories. Big-thinking companies don't view customers as individuals with individual concerns; they see them as part of "the mass market" or "knowledge workers" or "the Fortune 500." The consumer is viewed as "a male between the ages of 35 and 50, with an income over $50,000, married, two and a half children, and spends $150 per year on dog food."

These classifications, part of the American marketing practice for so many years, make us lose track of the real customers and the real influences in the market. All technology-based markets start small, adapting to the needs of particular customers. Companies succeed only if they focus on the specific needs of specific customers. Large markets always grow from small seeds. The personal computer began as a machine targeted at home hobbyists; only later did it grow into a billion-dollar business. The supercomputer began as a specialized machine for a few esoteric scientific calculations; now it is spreading to tens of thousands of applications in a wide variety of fields, from engineering to financial market analysis.

"Big thinking" also leads to bureaucratic and organizational problems. Companies that think big tend to split their organizations into "manageable" pieces. The organization then spends more time focusing on *how* to get things done than on *what* needs to get done. To succeed, you must create paths that allow ideas and individuals to cut through the bureaucracy. Top management must set the example. Top executives should spend time talking with and about individual customers, opening new paths through the bureaucracy, and rewarding individual initiative.

Until recently, thinking big was seen as a virtue, not a vice. Big companies could take advantage of economies of scale to drive small innovators out of the business. But in the 1980s something went wrong. Startup companies like Compaq and Sun defied the conventional wisdom. Now, small innovative companies have their goliath competitors on the ropes. Economies of scale can still be an advantage, but there are now a lot of footnotes. Footnotes that read: you must also be smarter than the other guy, act faster than the other guy, own market niches and build on them, build strong relationships with the infrastructure, and act like a small close-knit organization even as you start to get bigger.

■ **Try things.** No one has all the answers. No one thought about the microprocessor before 1972 or the personal computer before 1975 or the minisupercomputer before 1984. If anyone was thinking about these things, they didn't make them happen. How often do we have ideas that we let sit on the back burner or in the lab, only to watch someone else implement the ideas?

In facing an uncertain future, you have to experiment. You have to take risks. Success in the future will require risk-taking now. Of course, pioneers can sometimes end up with arrows in their backs. Failures will occur. But companies, like individuals, must learn to see failures not as defeats but as learning experiences. Failure can hurt in the short term, but companies can bounce back stronger and wiser—as long as they continue to take care of their customers and nurture their infrastructure relationships. Witness Apple's Lisa computer, Digital's personal computer effort, and Compaq's Telecompaq. All were failures, but all were also valuable learning experiences.

■ **Mix up your development groups.** Too often, familiarity breeds stagnation. When the same development group works together on one project after another, the group tends to lose its innovative edge.

Gene Amdahl has particularly strong views on this point. When a development group continues to work together, Amdahl explains, each member of the group tends to specialize. Each person handles the same part of the project each time. On the surface, that seems like a good idea: each person builds up expertise in a particular area.

But Amdahl sees it differently. He believes that people with "expertise" are dangerous to innovation. Once they get experience, they continue to do things the same way, time after time. They never try anything new. As soon as people get a particular assignment, says Amdahl, "they don the mantle of expertise. That is how they establish their positions in the [company's] society. I think we suffer from that very much in Silicon Valley." As a result, says Amdahl, "expertise masks innovation."

The problem is exacerbated by group dynamics. A group becomes a type of social organ. As members of the group build up areas of expertise, each person tries to avoid doing anything that would impinge on anyone else's area. At the same time, each person tries not to change his or her own part of the project too drastically from previous projects, since others in the group might be depending on certain things being the same as before. As a result, says Amdahl, "you end up with a [product] structure that's very similar to what you had before, except people have expended an exorbitant amount to try to make that [same] structure perform at a higher rate. And it's deadly."

What's the solution? Amdahl suggests rotating people among development groups: "You may take your most creative guy from the previous team, but you put him with new people."

■ **Encourage openness.** Many "crazy" ideas are just that—crazy. But some crazy ideas turn into innovations. If a company's environment discourages people from bringing up crazy ideas, the company might be more efficient in the short term. But it almost certainly will be less innovative in the long run.

Most innovative companies work hard to create an open, uncritical atmosphere. Compaq's Canion believes openness serves two purposes: it enhances creativity while also decreasing the risk of a major failure. He explains: "Openness means that when you see a problem, you're expected to flag it, to call it out, or at least to question that maybe there's a problem. Some cultures encourage people not to speak up because they don't want to hurt somebody's feelings. That leads to problems getting bigger and bigger before anybody really says something about them."

At the same time, says Canion, an open culture is also "a very good environment to be in to feel free to create. You're not worried about what you're going to say. You can say something that sounds dumb, perhaps, without fear of somebody criticizing you or somehow penalizing you. So the open environment has lots of implications to it. It's a free environment to create in."

■ **Create an environment, not a plan.** As I talked to successful innovators, two observations came to the surface. First, the atmosphere for innovation and creativity is driven from the top. Second, innovation itself comes from individuals or small groups somewhere in the bowels of the company.

Successful innovation is almost always a bottom-up process. Innovation can't be imposed from above; it must bubble up from below. So what is the proper role for top management? Management should focus on creating an environment for innovation, not creating a specific plan for innovation.

Digital has turned this strategy into a reality. CEO Ken Olsen provides the vision and the challenge, and he lets his people take it from there. "It isn't a matter of the people on top trying to define every little piece and push it down to the lower levels," explains Digital engineer Brian Fitzgerald. "It's the lower levels of folks who come up with the ideas that make the big vision come true."

❏❏ There's Still a Lot to Do

After two decades of rapid innovation in the computer business, there is still no end in sight. "The neat thing about computers is that there's still a lot to do," says Sun's McNealy. "I cannot walk into the office in the morning and say, 'Hi, computer,' and have it answer, 'Hey, how's it going?' I can't ask it, 'So what's my schedule today?' and have it tell me what my schedule is today. I can't expect it to answer when I ask it, 'So tell me what happened to the Dow Jones today. How's our stock doing? What are the analysts saying?' We can't do that yet. But we will.

"Some day I'll be able to walk in and ask, 'What happened in the news that's interesting to me today?' It'll answer, 'Blah blah blah.' I'll say, 'Well, that's not interesting!' It'll say, 'Sorry.' We'll literally be able to converse and we'll have access to all this database. It'll have AI capabilities; it'll be totally voice-interactive. If somebody else walks in and goes, 'Hey, computer, what's my schedule today?' it'll go, 'Sorry, you're not Scott McNealy. I won't tell you his schedule.' "

McNealy continues: "Until we get to the point in the computer and information age where the machine can do that, I'm convinced there are lots of areas where we can spend 13 to 14 percent of our sales dollars in creating new technology, in new ways of doing things like window systems, mouse interfaces, bit-mapped graphics, distributive computing, networking, AI capabilities."

Who, though, is going to do all of this innovating? And who is going to reap the rewards? Today, small challenger companies are the hotbeds of innovation. At these companies, there is a preoccupation with technology and innovation. As a result, challenger companies are churning out a seemingly endless flow of innovative products—and chipping away at IBM's once-dominant position in the computer industry.

But what will happen as challenger companies continue to thrive and grow? As they become larger, will challenger companies fall into the same traps as IBM? Will they be replaced by a new flock of challengers? As they grow, challenger companies will need to find new ways to build and maintain stimulating environments. In short, they will need to create new ways of doing business. That is the challenge for innovators of tomorrow.

> *"Now for the evidence," said the king, "and then the sentence."*
> *"No!" said the queen, "first the sentence, then the evidence."*
> *"Nonsense!" cried Alice, so loudly that everybody jumped, "the idea of having the sentence first!"*
>
> Lewis Carroll, *Alice in Wonderland*

◻ 6

Image

■■ OVER THE YEARS, IBM's image has been one of its great strengths. Often, the IBM image has conflicted with reality, but that hardly matters. If an image is strong enough, it can transcend reality. In effect, the image becomes the reality.

The power of IBM's image really hit home while I was talking to a fellow traveler a few years ago. He was the general manager of a rather large Fortune 500 company, and we had met on a plane. He mentioned that he always insisted that his company buy IBM equipment. When I asked him why, he replied, "Because I want all my computers to talk to one another." I asked him what made him think IBM's products all talked to one another. He paused for a minute and said, "Well, I *believe* they do."

I asked: "Do you know what *belief* is?" He shook his head. "It's a religion!" I said. "And religion is not based on rationality." As we went on talking, I said, "You can buy IBM computers for other reasons, but not based on the fact that they talk to each other."

The general trend toward networks and connectivity—and Digital Equipment's success in exploiting that trend—has long since exposed the reality of IBM's weakness in connectivity among its computers. But in his misconceptions about IBM, my fellow traveler made an important point about IBM. Despite reality, IBM's overpowering position, bolstered by an expertly maintained image, has the power to convince customers that IBM reigns supreme in all areas of computing.

Positioning has the ability to create perceptions that defy facts. These perceptions can be sustained over long periods of time, despite subsequent changes in reality. The Apple II computer created such a strong position that Apple Computer was able to maintain a leading position in the personal computer industry *despite* the failures of the Apple III and Lisa computers and *despite* the nearly ten-year lag between the Apple II and the Macintosh. The same is true for many other products, including Lotus's 1-2-3, Microsoft's PC/MS-DOS, Intel's 8080, HP's 3000, and IBM's 360 and 370. These products, by opening up new markets, created such strong images for their companies that the images remained intact even after other companies entered the markets with stronger product offerings.

The relationship between image and reality is a complex one. A company's image is initially based on reality: the performance of the company's products, the company's relationships with investors and customers, the financial strength of the company. Real value is a necessary (though not sufficient) prerequisite for building a strong image. Images depend on the word-of-mouth of early customers, the acceptance by systems integrators, the opinions of industry analysts, and so forth.

But images, while initially based on reality, can later transcend reality. As images take root, they become part of a customer's system of feelings and beliefs. With time, images can move far away from

Reprinted with permission, Malcolm Hancock, © 1988.

reality—but they are still a powerful force in the market. If a company builds a strong image early, it can often sustain its good reputation even in the face of product failures. On the other hand, a poor or weakened image can create doubts about a company, even one that offers cutting-edge products.

Over the years, IBM has often used image to its advantage. The computer giant has carefully cultivated a larger-than-life image, establishing itself (in the minds of customers) as a "national asset." This image influences the thinking of customers and competitors alike. Most important, it has enabled IBM to pursue a strategy based on FUD—Fear, Uncertainty, and Doubt. As I discussed in Chapter 2, FUD is the competitive sales tactic first attributed to IBM's attempt to stop the launch of Amdahl. The tactics are aimed at establishing major business and product concerns about competing companies in the minds of the potential buyers.

In this chapter, I begin by describing how IBM built its powerful image and used it as part of its FUD strategy. Then, I discuss how several challenger companies have battled against FUD and won— and have gone on to build strong images of their own.

❏❏ The IBM Image

IBM's image is unlike that of any other company. IBM is very much concerned with being perceived not only as a company, but as a national resource. It has tried to position its products as essential ingredients for national economic growth. According to IBM's carefully cultivated image, American business would grind to a halt without IBM products.

At the same time, IBM has tried to position itself as an important element in the social fabric of the country. While smaller companies focus most of their efforts on products, IBM spends a great deal of time portraying itself as a caring corporate citizen. It wants to be perceived as a company interested in everyone, right down to the proverbial "little guy." IBM wants to convince people that it is a truly benevolent giant, worthy of our trust and even our gratitude. As part of that effort, IBM contributes more than $100 million annually to public-service projects involving the arts, education, health, and welfare.

Joe Zemke, president of Amdahl and a former IBMer, says that IBM carefully plans its corporate image building. He points to a recent television advertisement as an example. The commercial focuses on *Writing to Read*, an IBM educational product that helps children learn to write and read. "I saw the ad this morning," says Zemke. "A schoolteacher says, 'It's the best thing I've ever seen in all the world. . .' What great advertising! They're very conscious of their image. Their whole advertising strategy is to create in the mind of the power brokers of America that IBM is a national asset. That was a strategy, and they've done a superb job."

IBM's image as a national asset became increasingly important as the federal government's thirteen-year antitrust suit against IBM dragged on. Through its advertising and lobbying, IBM tried to convince members of Congress and other government officials that IBM is a national asset. The message was clear: it would be a tragedy to break up—or even constrain—a national asset.

IBM's image building is probably one factor that helped convince government prosecutors to drop their antitrust suit as "without merit." Not surprisingly, the image building has also profoundly influenced customer thinking. For years, IBM's image has played a major role in persuading customers that only IBM could provide solutions to their computing needs.

IBM has built its image, in part, by controlling the flow of information to the media. It is a rare occasion and considered quite a coup for a journalist to get an interview with a high-ranking IBM executive, or for an industry conference to get an IBM executive to sit on a discussion panel. There are a few exceptions. Don Estridge, who founded IBM's personal computer operations, was a star on the industry circuit. But most IBM executives are consistently inaccessible.

The situation at the challenger companies is quite different. The challengers generally operate in a fishbowl. Journalists frequently interview executives from challenger companies for observations and analysis. Newspapers report problems and accomplishments at challenger companies almost daily. Indeed, company plans are often widely known before they are implemented. In the early days at Apple, we used to joke that it was "a ship that leaked from the top."

Despite (or perhaps because of) IBM's policy of limited access, many journalists see the computing world through Blue-tinted glasses. Richard Dalton, columnist for *Personal Computing* magazine, describes IBM like this: "They're in such a position of power, of influence, of control. At this point they would have to act at least neglectfully to put themselves in a position of degenerating."

Dalton, like many other analysts, credits IBM with eternal strength, thus perpetuating that image in the mind of the public. "You kind of measure this in celestial terms. There are waxings and wanings, but it isn't like the moon goes away. It comes up every night. Sometimes you just don't see as much of it."

❏❏ The Power of FUD

Overcoming IBM's image can be a formidable barrier for potential competitors. Consider the case of Gene Amdahl. In 1970, when Amdahl decided to start a company to compete against IBM in the mainframe market, he had trouble getting funding from venture capitalists. IBM was perceived as such an omnipotent force that Amdahl's business plan was rejected repeatedly before he finally got funding.

"At that time, the venture capitalists that we talked to were really not ready to be convinced that it was possible to compete head-on with IBM," recalls Amdahl. "Basically, they felt that IBM had machines on the shelf it could bring out at a moment's notice that would far exceed anything anyone else could do. I never saw such a shelf in IBM. There never was a shelf that I ever knew of. Every machine that was put out on the product line was the best that they could do."

The venture capitalists repeatedly pointed out how previous efforts to compete against IBM had always failed. RCA was the prototypical example. After spending about $500 million in its effort to compete against IBM, RCA estimated that it would have to spend another $500 million before turning a profit. At that point, RCA decided it would be better to cut its losses.

To the venture capitalists, that was a clear signal that it was not possible to compete head-on with IBM. Their argument was: If the big guys couldn't succeed against IBM with a billion dollars, how could upstart Amdahl do it with a few tens of millions? Amdahl logically debated the question, trying to convince doubting investors that he could indeed succeed. Amdahl argued that his approach was much different from RCA's. For one thing, Amdahl planned to build an *IBM-compatible* mainframe—that is, a mainframe that would support the same software and peripherals as an IBM mainframe. "I had a devilish time trying to convince investors that RCA didn't try to make their computer compatible; in fact, they tried to make it incompatible," says Amdahl. "They tried to make it incompatible in such a way that the user could fairly easily be brought on to the system, but when he got done using it a while, there was a barb on the fish hook—he couldn't get away without having to put in a great deal of effort. That was their strategy."

None of Amdahl's arguments seemed to help. The venture capitalists believed that it was impossible to compete head-on with IBM. They felt that IBM had some sort of magical power to make moves, at its whim, that would destroy competitors. Eventually, Amdahl did convince one venture capitalist—but only one.

Convincing investors was only the first hurdle. Convincing customers was just as difficult. Getting beyond IBM's established image, its overwhelmingly dominant position, was a mammoth task. "We made a survey before we sold any of our machines," says Amdahl. "Both in the U.S. and in Europe, we got nearly the same results. The results told us that customers wouldn't consider buying anything other than IBM unless you had price advantage of at least 15 percent or performance advantage of at least 20 percent. We also found that 30 percent of the customers polled would not consider, under any price differential, buying anything other than IBM."

Part of the reason for this customer loyalty was FUD—Fear, Uncertainty, and Doubt. The term FUD was coined in the early 1970s, when Gene Amdahl's new mainframes became a threat to IBM's machines. Gene Amdahl remembers: "I'm given credit for coining that name, but I didn't. It was actually coined by the group inside IBM that was put together to produce a strategy to stop Amdahl. One of my friends from inside IBM informed me about the strategy. It was a very effective approach."

According to Amdahl, IBM glorified its new product to scare

away potential Amdahl customers. IBM salespeople told customers what IBM was going to do in very general terms—and it sounded as if IBM's new products were going to be remarkably different. In fact, they weren't. "We could show that they couldn't do what they said they were going to do," says Amdahl. Through some technical detective work, Amdahl was able to deduce what IBM was actually going to do. Amdahl explains: "We went through and figured out the main functions in the operating system that they could actually have put in there. We had a bright young guy, must have been maybe twenty-five, and he went through it and he decided that there were really about fourteen or fifteen things that they could do. When they came out with the new product, it did thirteen new things, and all but two of them were ones that we had identified. We were able to announce six weeks later the same functions, which must have been quite a blow to the IBM FUD program. But they still maintained FUD from then on, nonetheless."

❏❏ The Fading of FUD

Courtesy of
Amdahl Corp.

The battle against FUD is much easier today than it was for Amdahl. Indeed, IBM's FUD tactics do not pack as much punch as they once did—and they will become even less effective in the future.

Educated investors are already pretty much immune to FUD. Industry stars like Gene Amdahl no longer need to convince venture capitalists of the feasibility of challenging IBM. Times have changed substantially. Since 1987 alone, venture capitalists have invested $1.2 billion in computer and computer peripheral companies—many of them competing directly with IBM. Since 1980, venture capitalists have invested $6.4 billion in the computer industry. These funds have launched not only well-known industry leaders like Compaq, Apollo, Convex, and Sun, but also thousands of others like Ardent and Alliant and Sequent, and lesser known startups with intriguing names like Synaptics and Saxby and Voysys.

Nor are customers as susceptible to FUD as they once were. There are at least three reasons for this fading of FUD. First of all, customers are much more educated than they once were. "The marketing power that IBM exhibited in the 1960s and 1970s made them synonymous with computers. Senior management was afraid to trust anyone other than IBM," explains Max Hopper, vice-president at American Airlines. "If you bought IBM, even if you failed, you didn't lose. But the minute you thought of buying non-IBM in those days, the IBM salesman or senior management was camping on your chairman's doorstep warning of all kinds of impending doom. That marketing tactic doesn't work any more. In most companies now, the people responsible for data processing and purchasing have smartened up. They've gotten enough sense to, in effect, push down that IBM marketing edge."

A second factor is the rise of systems integrators—companies that put together "mixed" computer systems, using the best parts from many different vendors. "If you go to a systems integrator, you're not going to buy an IBM PC, you're going to buy the best product," says Dave Martin of National Advanced Systems. "To compete against IBM, to fight all that FUD, a company used to need an enormously better product just to get the customer to consider it. Now, there's still FUD there, but it's been reduced. When a customer goes to a systems integrator, there's zero FUD. In fact, it's reverse FUD. If the systems integrator buys from IBM, he knows he has to take money out of his own pocket; he knows IBM is not really going to support him because IBM really wants the business itself."

A final factor contributing to the fading of FUD is the change in IBM's image itself. Pockmarks are now appearing on IBM's once-unblemished image. DEC's success in winning sales in traditional IBM territory has been particularly damaging. The fact that a competitor has even a chance to beat IBM on its own turf tarnishes Big Blue's image. That DEC is consistently beating IBM is devastating to IBM's image. As Dave Martin of National Advanced Systems explains: "The DEC phenomenon said: 'IBM didn't do a good job for you, major customers of the world. DEC did.' It has hurt IBM terribly in terms of its image."

The deterioration of IBM's image is perhaps best symbolized by the debacle of IBM's PCjr. The PCjr, introduced in late 1983, was IBM's attempt to sell computers into schools and homes. IBM's PC was already a big success in the office. IBM wanted to repeat that success in schools and homes: it wanted to drive out the competition and gain control of these potentially huge markets.

In typical FUD fashion, rumors about the PCjr began circulating months before the actual introduction. Everyone in the industry was talking about the Peanut, IBM's code name for the PCjr. When finally introduced in November 1983, the PCjr became the most hyped machine in the history of the computer industry—or any industry. Expectations were very high. Newspapers and magazines lavished praise on the PCjr, heralding the new machine as the "third coming" after the Apple II and the IBM PC. Here's a sampling of press clippings from November 1983:

> Newsweek: *The PCjr, despite its guise and price tag, is not a humble little home computer; it is a powerful machine that may reshape the computer market for years to come—and in favor of Big Blue.*

> Washington Post: *The PCjr, as the machines are called, signals the determination of the world's largest computer company to achieve the same preeminent position in the home as it currently enjoys in business.*

> Business Week: *Because it is the world's largest computer company, IBM will undoubtedly convince many up-to-now uncertain consumers that computers actually have a place in the home.*

Everyone assumed that the PCjr would be a huge success. No one ever considered that it might fail. After all, didn't IBM machines always succeed, even dominate, in the marketplace?

But a funny thing happened on the way to the market: customers didn't buy the hype and they didn't buy the computer. IBM got off on the wrong foot by introducing the computer in November, three months before it was actually ready to ship. By introducing the PCjr in November, IBM hoped to slow the Christmas sales of competing computers. But the strategy backfired. Consumers didn't want to wait until February, so they bought other computers for Christmas. In fact, Apple had the best holiday season in its history. Dealers, meanwhile, were furious with IBM for trying (albeit unsuccessfully) to cut into their holiday sales, since they depended on holiday business for a large portion of their annual sales.

As people began to look at the PCjr more closely, it no longer looked like a magic machine. The machine's keyboard was particularly bad. The keyboard had widely separated keys that looked somewhat like "chiclets." The legends were printed not on the keys themselves, but in between, forcing users to tilt the keyboard to actually read the legends. Worse, the keys were awkward to use and did not have the "quality feel" of the keys on the Apple II and the IBM PC. A review in *Byte* magazine put it like this: "The keys themselves have a mushy feel, most unpleasant for touch typing. I get the impression that IBM designed this keyboard for children, and perhaps that's why I don't like it."

The price of the PCjr also raised questions. IBM had given the PCjr a premium price to reflect the premium brand name. But the product didn't live up to the name. Even if it had, consumers might have resisted the high price. The dynamics of the highly competitive market just weren't right for a premium-price product.

IBM tried to counter the initial consumer resistance with a major promotional campaign. In the summer of 1984, Charlie Pankenier, director of communications for IBM Entry Systems Division, brought all of IBM's dealers to Dallas for a corporate pep rally. Pankenier unveiled the new promotional program to a ballroom full of dealers. At the core of the program was a direct-mail and advertising campaign aimed to achieve more than 2.5 billion impressions (number of viewings) during the next 120 days. The campaign was targeted at the PCjr's "primary prospects," whom Pankenier described as people with "high income levels, college education, and children living at home." The campaign hoped to reach 98 percent of the U.S. population, with each person exposed to an average of 30 PCjr promotional messages.

The advertising campaign included placements in 160 newspapers and 80 magazines, including seventeen pages of ads in the *Wall Street Journal* and twelve-page inserts in *Time, Newsweek, Sports Illustrated,* and *Readers Digest.* The *Readers Digest* insert was specifically targeted at "people who hadn't even thought about buying a PC yet." Television spots were aired on every "Monday Night Football" game and on broadcasts of every game during the baseball Championship Series and the World Series.

IBM also announced a special joint promotion with consumer giant Procter & Gamble. The direct-mail campaign, aimed to reach 45 million households, offered consumers a "$100 off" coupon toward the purchase of an IBM PCjr in exchange for twenty-five proof-of-purchase coupons from participating P&G brands. The campaign also included a sweepstakes with 250 PCs and 750 PCjrs as prizes.

Finally, IBM slashed the prices on its PCjr system. A full system with color monitor, priced at $1,698 at the product introduction, dropped to $1,269, then to $999. In addition, IBM offered a $250 rebate to dealers, and it encouraged the dealers to pass the rebate along to consumers.

IBM's consumer-style promotional campaign flopped. IBM reportedly spent $100 million promoting the PCjr, but the money was wasted. The damage had already been done. One large computer dealer told me at the time that he could not even give the machines away. In March 1985, IBM pulled the plug, announcing that it would halt production of the PCjr. *Time* magazine reported on the demise of the PCjr: "The withdrawal of the computer giant from the home market marks a rare and embarrassing chapter of failure in IBM's long record of industry dominance."

The damage to IBM's image was enormous. After the PCjr, IBM was perceived as vulnerable. As powerful as IBM was, it could not force an unwanted product on the customer. Shortly after the PCjr, IBM had several other product disappointments, and its image continued to deteriorate. "Companies have been misled," says Compaq president Rod Canion. "IBM said certain new products were going to be good and they weren't. So IBM's credibility has suffered. People still respect the IBM name, but they have to prove themselves now."

Customers, dealers, and reporters now look at new IBM products with a much more skeptical eye. This new skepticism was apparent in the reaction to IBM's Personal System 2 family of computers, introduced in 1987. "With IBM's major new products coming in April, the expectation around the market was a slowdown in the overall industry and particularly a slowdown for Compaq," says Compaq's Canion. "But March was our strongest month in sales ever. We set a new record in March!"

After the announcement of the PS/2, skepticism continued, particularly in regard to the Micro Channel Architecture, the new communications channel that IBM built into the new computers. According to IBM, the Micro Channel moves data through the computer much more quickly than previous technologies. IBM is keeping certain details about the Micro Channel proprietary, hoping that the Micro Channel will provide it with a technological advantage over companies making PS/2 imitations.

Losing Credibility

IBM's PR machine was hard at work at COMDEX trying to dispel the disbelief about PS/2. But we're still skeptical. Yes, the company says it has shipped a million PS/2s and presented us with the chemistry professor who bought machine number 1 million. So? It still won't tell us how many of these machines have been sold. Shipped just means they aren't sitting around inside IBM somewhere. Despite the nationwide "Thanks a Million!" advertising campaign, we still think the real number of PS/2 machines sold (real meaning those with a Micro Channel) is less than a quarter million.

Source: Technologic Computer Letter, November 9, 1987.

Had IBM announced a "magic ingredient" like the Micro Channel ten years ago, or even five years ago, the media would have accepted it at face value as a major technological breakthrough. Customers would have rushed out to buy a computer with new Micro Channel capabilities. But the computer industry works differently these days. This time, the media checked with the industry infrastructure more carefully, and reporters found that not everyone considered the Micro Channel such a breakthrough. In fact, quite a few people viewed the Micro Channel as an IBM marketing ploy, designed to charge premium prices and to discourage customers from buying IBM clones.

Indeed, the *Wall Street Journal* ran a front-page article full of skeptical reaction to the Micro Channel. Here's a sampling:

The Micro Channel isn't mere technology. It is IBM's mystery ingredient: an electronic version of the "MFP" in Colgate toothpaste or the "Chlorinol" inside Comet cleanser, intended to give IBM's PC extra zing in a crowd of look-alikes. As such, the Micro Channel is one of IBM's most audacious marketing gambits ever. . . .

A lot of smart computer people have been trying to find out what goes on inside a PS/2. And some of the industry's top engineers say that after spending months poring over the Micro Channel, they haven't found much that customers really need or that IBM rivals can't match.

The article quotes one industry source as saying: "The emperor's wearing a bathing suit." And people are beginning to notice. The IBM image just isn't what it used to be. IBM's image no longer defines reality.

❏❏ Building an Image

IBM's failure with the PCjr provides important lessons on how *not* to build an image. In marketing the PCjr, IBM adopted a consumer-product approach: it promoted the PCjr as if it were a new soap or detergent. But technology-based businesses are fundamentally different from consumer businesses. In the world of technology, you can't build an image with television commercials and money-back coupons.

Consider a person buying a can of peaches at a supermarket. The customer might buy a brand-name product, because he *believes* that the well-advertised brand-name product contains higher quality peaches than the generic brand. A "peach expert" might know that the generic product uses exactly the same peaches as the brand-name product. But how many of us have access to information from peach experts?

The situation is quite different in technology businesses. Customers have access to more information about products, and they base their decisions on *real* differences between products. No one buys a computer without checking references, without talking to users and other knowledgeable sources. Customers need to be more knowledgeable about the computers they buy. Customers don't need to worry about compatibility between their breakfast cereal and their milk, but they do need to worry about compatibility among various types of computers and software and peripherals. Buying a computer involves a type of commitment that doesn't exist in consumer products.

There are many ways for computer customers to obtain knowledgeable sources of information. Newspapers and magazines are filled with reviews and information about various computer products. And the computer itself is helping make expert information more readily available: people looking to buy a computer can use computerized information services to get expert advice to inform their decisions.

As the computer industry fragments into narrow vertical markets, expert information is becoming even more readily available. Consider a software package designed specifically for lawyers. You can be sure that lawyers will talk with their friends and colleagues in the law community before buying the package. Word-of-mouth becomes a primary carrier of the market's confidence in a product.

As expertise becomes more necessary and more readily available, buying patterns change. Brand is the refuge of the insecure. People buy brands because of what they *believe* but do not necessarily *know* about the product. But as people become more knowledgeable, they no longer need to rely on brands. That is why an increasing number of major corporations are buying IBM clones. They know that the only difference is the price and the name.

In short, knowledge allows customers to go beyond brand. In the computer business, brands must reflect *real* differences in products. I have often said that public relations is dead. What I mean is that advertising and public relations do not create the market for computers: they are used to reinforce and reflect what is happening in the market. There must be tangible results to create and back up the image.

So technology companies need new ways to build images for themselves and their products. In today's technology-based industries, image is much closer to reality than ever before. Technology-based products can't be sold on the basis of false perceptions. Perception and reality are coming closer together. Image, in today's technology industries, is a reflection of reality. Image is built on good products and relationships. Without these, nothing else matters.

In the early days of Apple, employees often debated whether Apple should look and sound more like IBM. People who joined

Apple from companies with thirty or forty years of tradition often insisted that Apple, to be successful, would have to emulate IBM in style and manner. One young Apple manager once told me: "We've got to look like IBM. We've got to look big and professional, dress like business people, and wear ties like IBM."

That is precisely the wrong way to build an image. Image begins with the right product. It wasn't a change of clothes that opened the doors of corporate America to Apple; it was a change of products and a demonstration of market acceptance and financial stability.

The key to building a good image, then, is to build a good reality. In particular, image is influenced by the quality of a company's products, its investors, its users, and (perhaps most important) its financial performance.

❏ *Products*

The challengers who succeed today are very pragmatic about providing customers with the right products and solutions, and they are very pragmatic about their images. Successful challengers are technology-driven and solutions-driven; their images follow naturally from that orientation. Tandem's image, for instance, can be summed up as "doing a good job for our customers." Everything else follows. Compaq's image is a reflection of its reliable, dependable personal computers, and its unceasing dedication to its retail dealers.

Today's customers are naturally skeptical about new technology products. Vendors can overcome customer uncertainty only by consistently meeting or exceeding the expectations of the marketplace. That takes some discipline. In effect, it means being honest about what you can and will achieve, while projecting a healthy future for yourself.

❏ *Investors*

Startup companies begin at a big disadvantage in the image game. They must prove their long-term viability. John Rollwagen recalls the early days of Cray: "There has always been this major threat. It has been there from the beginning. When we first went public, it was

Burroughs that was going to eat our lunch because it had just introduced a high-end product. For a while it was even Goodyear because it had just announced a version of a supercomputer. Anybody bigger than we are who announces anything is already ahead of us. There's this fascination with bigness."

One way that a startup company can gain credibility is through its investors. I have often told startup companies that *who* invests in them is much more important than how much is invested. In other words, the *quality* of investors says a lot about the market's confidence in the company. Funding from a respected venture-capital group is worth more than a dozen prime-time commercials.

Similarly, Compaq gained credibility when it decided to join the New York Stock Exchange. It was one of the very few venture-backed startups in recent history to join the Exchange. That move helped position the company so that it was perceived as a more stable and financially strong computer company.

❑ *Relations with the Industry Infrastructure*

There are a lot of knowledgeable people between the vendor and the user. These people make up the industry infrastructure: dealers, value-added retailers, third-party software companies, systems integrators, industry analysts. The computer industry, despite its multibillion-dollar size, is like an extended family. Everyone knows everyone else, even if only as a distant cousin.

Connections within this extended family are crucial to success. A strong product is not enough. If you hope to succeed, you must develop a web of connections with the rest of the industry infrastructure. Relationships and connections can have a snowballing effect. Companies that have developed connections often get the first shot at new ideas. When a new software company or new communications company or new peripherals company readies its first product, it will first seek out some affiliation with one of the major players in the industry. Companies with connections will get the call.

Microsoft and Sun have probably exploited the industry infrastructure better than anyone else—and that is a big reason why those two companies are among the most respected and feared in the industry. Their connections and affiliation roots go deep into the industry. Microsoft chairman Bill Gates has the opportunity to see more of what goes on behind the scenes in the microcomputer industry than anyone else. He has relationships with all of the industry leaders, including IBM and Apple. Thus, he is positioned to take advantage of opportunities before others can.

The opinions of people in the industry infrastructure strongly influence a company's image. If people in the infrastructure endorse a new product, users are likely to follow. One way the infrastructure influences image is through the media. Journalists no longer accept press releases as fact. They interview key players in the industry infrastructure and report on how these players view the new product. If the infrastructure lines up behind a product, everyone is likely to view the product as a winner.

❑ *Initial Users*

Early users of a product are important in several ways. In the premarket stage, so-called *beta sites*—users that try out the product before it goes to market—can test, alter, and adapt the product so that it more closely fits the needs of the market. Through a process of interactive adaptation, the product evolves into its finished form.

Once the product is introduced, the beta customers serve a different role. A vendor is judged (at least indirectly) by the quality of its beta sites and other initial users. If the initial users have quality images, the product will gain a quality image too. The initial users can also help spread the word about the new product. Such word-of-mouth referrals are the most effective form of advertising.

❑ *Financial Performance*

Once a company's image is established, the image can be reinforced—or reversed—by the company's financial performance. This is particularly true of high-technology companies. Few of us are concerned about the financial condition of the consumer-products companies that make the products we buy. When we buy cereal or toothpaste, we don't ask the seller about the financial condition of the producer. Even during the 1970s, when the U.S. auto industry was in terrible shape, consumers continued to buy American cars.

But the situation is quite different with computer purchases. If people think a computer vendor is in financial trouble, they will stop buying the company's products because they will begin to question the company's viability. Computer customers are especially cautious because they have long-term expectations. All computer purchases involve planning for the future. When people buy computers, they ask questions like: Will this computer continue to serve my needs as we grow? Will the supplier keep it state-of-the-art? Will I be able to upgrade the product, acquire enhancements, rely on service and support over the long term?

A company's financial performance strongly influences a customer's "comfort level" in dealing with the company. For example, Digital's financial success in 1987 reinforced customer confidence in the company. The result was a positive feedback loop: as customer confidence rose, Digital's financial performance continued to improve, leading to further increases in customer confidence, and so on.

❏❏ Case Studies

Tandem Computer and Apple Computer have followed these guidelines well. By focusing on products and relationships, not promotions and gimmicks, both companies have built strong images for themselves and for their products.

❏ *Tandem*

Tandem founder Jimmy Treybig exemplifies the triumph of substance over style. Treybig always looks like he's working hard: he runs around his company with his shirt hanging out. But Treybig says his customers don't worry about his style of dress. They don't care that he doesn't look like an IBM executive. What they want is a machine that works, a machine that gets the job done. Tandem has always provided that.

Tandem has also built an impressive reference structure. Tandem's first installation was Citibank in New York City. That installation gave the company a healthy start. People saw Citibank on Tandem's customer list and figured that Tandem must be a serious startup.

Significantly, Tandem spent a lot of engineering time with Citibank and other early customers. Tandem didn't just sell computers and ask for references: it worked closely with the early customers, listening to their suggestions and making sure they were satisfied with their systems.

Tandem focused on building an image of reliability. Tandem's computers were the first "fault-tolerant" computers in the industry, guaranteed never to shut down. To solidify its image of reliability, Tandem set up a special system in its home office so that it could monitor all of its customers' computers on a real-time basis. In many cases, Tandem was able to flag customer problems before the customers even knew about them.

Tandem's image was challenged in 1985, when other vendors began to offer fault tolerance as a standard feature of their transaction-processing machines. So Tandem worked on shifting its image. Now it wanted to develop its image as a company that "got the job done" for its customers. Rather than simply advertising this new slogan, Tandem worked hard to make it a reality. When Tandem installed a new machine, it focused on helping the customer achieve quick results. Tandem was totally dedicated to making its products work in customer environments. After Tandem had gathered a few success stories, it spread the word through its sales organization, at industry conferences, and through its user groups. And Tandem executives used every reference and contact they knew to get in the door of potential customers.

This strategy worked because Tandem's image is a reflection of reality: the company really *is* dedicated to customer satisfaction. Madison Avenue could never create such a strong image.

❏ Apple

In its early days, Apple was dealing in a totally new market. So it played by a somewhat different set of rules: it let its image get ahead of reality. At Apple, the vision became the reality. Steve Jobs talked more about the vision than about the products. The goal was to convince people that there really was a future in this new market.

Many people cite Apple as a company that used advertising and promotion to build its image. But that is not so. Underlying Apple's image are Apple's unique and innovative products. Apple computers have won incredibly strong support from the industry infrastructure. Within a couple of years of the introduction of the Apple II, more than two hundred companies were developing software and peripherals for the machine. They didn't support the Apple II because of Apple's advertising and promotion; they supported it because of the innovative nature of the machine. The support of the industry infrastructure, in turn, created new markets for the computer, and reinforced Apple's image. Without the support of the infrastructure, Apple's advertising and promotion would have accomplished nothing.

With the Macintosh, Apple started with an image problem. The early Macintosh was compared directly with the IBM PC, and on several counts it came up lacking: there wasn't enough business software; the computer itself didn't have enough memory; there were no "slots" to expand the computer; the company didn't articulate a clear product direction. And when Apple's earnings declined in 1985, potential customers became concerned about the company's future.

Apple began improving the Mac's image by articulating a growth path for the machine. Apple had to convince third-party software companies that future Macs would have sufficient memory for serious programs. And Apple had to convince business users that future Macs could be expanded to run programs intended for the IBM. Businesses might never use the Mac in this way, but the link to IBM made them feel more secure.

As Apple began to roll out these enhancements to the Mac, the Mac's image (and Apple's image) began to change for the better. But the biggest change came in the first quarter of 1986, when Apple announced an improved Macintosh and a dramatic rise in earnings. Immediately, the industry and the press proclaimed that Apple was back—even though the company's sales were flat and many new products were still in development. The important thing was that Apple had regained the confidence of the market, so Apple's image shined.

□□ A New Image for Image

For many corporate executives, image has a bad image. Executives are often schizophrenic about images. Naturally, every executive wants his or her company to have the best possible image. On the other hand, executives often show disdain for the word *image*. The term often implies a façade, or a publicly reflected view that is not real in substance.

To succeed in today's technology-based industries, companies need to develop a new image for image. Image is just a reflection of reality. Building a better image depends more on *process* than on particular *events*. Companies should think less about promotional events, and more about the process of developing strong relationships and educating the industry infrastructure.

Our audience made us successful.

George Burns

❏ 7

Customers

■■ I RECENTLY HAD DINNER with Apple co-founder Steve Jobs. Steve's new company, called NeXT, was about to launch its first product, a powerful desktop computer. Steve said that the most valuable thing he had learned since leaving Apple was the importance of talking to computer users. I asked Steve what had changed as a result of talking to users, and he replied: "Everything, and in every way." He then explained: "When we first went out and asked potential users what they wanted, we listened hard, came back to the lab, and said to ourselves: 'Something must be wrong. What they say they want isn't that hard to build.' Then we went back and asked them: 'What do you want to *do* with your computer?' The answers were completely different, and the solutions were much tougher to implement."

Of course, talking to customers is hardly a radical idea. Every businessperson knows the importance of talking to customers. But many companies don't listen carefully enough and they don't *act* on what they hear. They don't adapt their products in response to what the customers say.

IBM is the prime example. IBM has a wonderful sales force. IBM salespeople build strong relationships with their customers, and they often know their customers' business better than the customers know it themselves. But the central purpose of this effort is to sell computers, not to improve the product. I call IBM a *sales-driven* company. In interacting with customers, its chief goal is to push more computers through the door. A technology challenge doesn't seem to stimulate IBM the way a sales challenge does.

Successful IBM challengers, by contrast, are usually *market-driven* companies. They are willing to be adaptive. They continually adapt their products, improve on them, enhance their solution-making capability, and work with other companies to provide more complete solutions.

Many computer companies are driven by technology. Challengers unravel the edges of science and technology by developing better semiconductors, magnetic media, optics, artificial intelligence techniques, and so on. But better technology isn't enough. Market-driven companies keep one foot in technology, so that they know the potential, and one foot in the market, so that they know the opportunities.

Market-driven challengers start by understanding the competitive environments of their customers—and then find ways of adapting their technologies and capabilities to help customers become more competitive in their businesses. They find ways to adapt their computers and services to the needs of the customer.

In many cases, the challengers cannot meet customer needs by themselves. They cannot offer all the peripherals, software, networking, and support. So they build relationships with other companies—and design products that other companies can easily adapt and extend. The Apple II computer thrived not so much because Apple adapted it to the needs of customers as because third-party companies kept adapting it. Adaptability is the name of the game. The challengers might not have salespeople sitting on every customer's doorstep every day, and they might not be as good at hand-holding as IBM. But they are far more adaptable and willing to make changes to meet customers' needs.

To a certain extent, IBM deserves much of the credit for making the challengers so market-driven. Because IBM dominated the large-volume accounts, challenger companies initially had a difficult time gaining a foothold. So the challengers *had* to be more flexible and accommodating. As the new kids on the block, the challengers couldn't afford to be arrogant.

In this chapter, I describe the market-driven strategies of several of the successful challengers. But first, I take a look at the opposite approach: the sales-driven strategy of IBM.

❑❑ The Sales-Driven Approach

IBM's sales-driven approach to marketing dates back all the way to 1914, when Thomas J. Watson Sr. became general manager at the company. Watson, a former salesman himself, paid special attention to the company's sales operation. He was once quoted as saying, "Collecting salesmen is my hobby." Watson elevated the status of salesmen within the company, awarding them high salaries and generous bonuses, and he expected the highest possible return for his investment. Salesmen were (and still are) expected to know their customers personally, to know their customers' companies inside out, and to understand fully the workings of customers' industries.

"The sales guys were given the responsibility of understanding their customers' businesses," says Bob Miller of MIPS. "If you brought the guy in who was the account manager for Ford, he'd better recite chapter and verse what the automobile business is all about, not just be able to say, 'I took Joe Blow from Ford out for a beer and he's going to sign a purchase agreement.' That guy had to know everything from how Ford kept its ledger to how Ford did its forecasting. I mean, he was responsible for understanding that business."

IBM also expected its salespeople to develop strong personal ties with influential people at customer companies. "If I still worked at IBM," says Amdahl's Joe Zemke, "I would know people at General Motors and other places that I worked during my eighteen-year career. If there was a tough marketing situation, they'd trot me back in there. I'd be dealing with guys that I taught to program in 1964, that I went to baseball games with in 1966." Strong bonds form through such shared experiences. "We stayed up all night installing the first manufacturing control system in the Chevrolet Gear and Axle plant in Detroit," says Zemke. "That's pretty powerful stuff. And they've been doing it for a long time."

By working with their customers, and learning their customers' businesses, IBM salespeople evolved into very good businesspeople. Indeed, the best salespeople at IBM kept moving up the hierarchy, becoming branch managers, then regional managers, then vice-presidents. The more aggressive the salesman, the more bonuses and elite opportunities he gained within the company. All six of IBM's presidents to date have risen through the sales ranks. So have many other top IBM executives. Zemke rattles off the names of great IBM salesmen the way schoolboys rattle off the names of baseball stars: "Akers is a great salesman. He is one of the best salesmen I've ever seen in my life. Terry Lautenbach, Ed Lucente, and Mike Armstrong are also very good salesmen."

IBM's salespeople know that the path to the top is through effective sales techniques. Their mission is to "get the order." Once that is accomplished, the next challenge is "keep the order." After that, "increase the order." To increase sales, IBM salespeople spend a considerable amount of time convincing existing customers that they need the next-generation computer.

For many years, IBM salespeople were much more concerned with increasing the order than with fending off competition. Bob Miller recalls what it was like when he joined IBM in 1966. "The 360 was just beginning to ship and there was no question that there was a very small world of competitors in that period," says Miller. "I used to say an IBM salesman's biggest problem wasn't competitors, his biggest problem was how he nurtured an account and got the account to buy more business."

When IBM salespeople talked with one another or gathered at sales meetings, they rarely talked about the competition. "It was basically, 'How do I get the MIS guy to put new programs in place so he's going to need new equipment?' That's my perspective of what was going on," recalls Miller. "There were a couple of situations, I'm sure, that were hotly contested. But in general, it was everybody operating out of their own installed base. IBM by that time had developed a much better installed base. The 360 went out like gangbusters on top of the 7090. The key was that they already owned the Fortune 500 by that time. So the whole name of the game was just convincing Fortune 500 MIS people how they needed more." The task was straightforward, since buying decisions were very localized. "I mean, you didn't even have to go deal outside the MIS guy," says Miller.

Even to this day, IBM pays the most attention to other computer companies when Fortune 500 accounts are at risk. "IBM doesn't get mad at anybody until they steal a Fortune 500 account," says Miller. "You can be the nicest guy and they don't care at all. You can sell products to one of their accounts all day long. But go in and steal Mobil Oil or Ford or General Motors, and all of a sudden the crosshairs come on you."

IBM's close ties with customers, and its emphasis on "increasing the order," has been largely successful in the past. IBM's personal-touch approach allowed IBM staff early access to their clients' opinions, doubts, and concerns. Thus, IBM got the first chance to solve customer problems—before competitors with potentially better solutions were even aware of them.

But IBM's sales-driven approach does not offer the best service to the customer in the long run. IBM often ends up selling the customer what it can *make* rather than what the customer *needs*. Indeed, IBM has often manipulated, rather than served, the customer. The top goal is always increasing the order, not providing better solutions to customer needs.

In many cases, IBM's sales-driven approach has kept state-of-the-art technology out of the hands of its customers. In the mainframe market, IBM has paced improvements not according to advances in technology, but according to financial calculations. The company orchestrated a series of incremental improvements that would maximize profits. IBM's foremost concern was IBM—not giving customers the most technologically advanced product for the price they paid. Because of its position in the industry, IBM set the standards and told customers what to expect and what was good value.

IBM is beginning to pay the price for its deep-seated sales mentality. "A lot of the companies that have competed successfully against IBM were run by technical people, and I think that's where IBM's flaw is," says Bob Miller. "Knowing the business is what made IBM successful, but it also opened the door to the innovators because salesmen are not out there to be strategic, innovative thinkers. They're out there to get orders."

Miller continues: "I used to say that a sales guy's idea of a major change in the product was to change the keyboard and whatever he heard from the customer last. So salesmen are into small, incremental improvements, not the over-the-horizon. How many sales guys have you ever met who are going to think through what the computer industry could look like in five years? No, they gauge more what people are buying at present and what the last competitor knocked them off with."

What's more, salespeople generally don't have the sense of commitment to products and technologies that technical people have. "That's the point about sales guys: sales guys don't have the same emotional sense of ownership of things as engineering/scientific types," says Miller. "It's like I keep explaining to my assistant when he gets frustrated with the engineers. I say, 'That's part of the part number.' I say, 'If you write the specs of what makes an engineer versus what makes a salesman, that's what comes with the part number.' That used to be the frustration at IBM for [technical] people like me."

IBM's sales-driven approach worked fine in the era of the mainframe. The uses of computers, and the needs of customers, evolved only slowly in those days. But today's environment is very different. Today's computer customers don't want a sales pitch. They want a computer vendor that will adapt its technology to meet their needs.

❏❏ The Market-Driven Approach

IBM's challengers became market-driven almost by necessity. Anyone who hoped to compete with the all-powerful IBM could do so only by being extremely responsive to potential customer needs. They certainly could not take IBM's approach of forcing customers to adapt to whatever solution they made available. Instead, challengers came to potential customers hat in hand, saying, "Look, I have this new thing. I'd like to show you how it works." If the customer said he needed compatibility, the challenger responded: "Fine, we'll put a compatible board in it, or we'll provide a link or bridge to the IBM world."

This approach serves the customer better than just trying to sell and maintain orders. IBM salespeople are very customer-responsive, but they're most responsive when they think they're going to lose an order. They fly people in, give slick presentations, and grace clients with visits from persuasive IBM top management. In short, they do whatever it takes to save the order—and preserve the IBM sales ethic.

There's nothing wrong with being competitive, with fighting for every order and not giving up. But when a company fails to put its sales effort into the context of the customer's view of the world, the company risks losing its credibility in the long-term. Being customer-oriented may require some radical changes in both the product and the ways a company does business. Sales-driven companies tend to think in terms of changing the customer's mind to fit the product; market-driven companies change the product to fit into the customer's strategy.

To a certain extent, market-driven strategies are based on familiar ideas. "It's the old stuff, you know, watch your customers," says venture capitalist Floyd Kvamme. "Serve a real market and don't just try to rip off an IBM market. Serve a real customer." But few companies carry out these strategies well. Few companies are truly responsive to customers. They might listen, but they don't adapt and change.

Listening to customers is just a start. There are other important elements to a market-driven approach. Here's a list of guidelines.

■ **Keep one foot in the technology, one foot in the market.** Management must know the potential of the technology and also see the opportunities of the marketplace. Large companies like Xerox and IBM often fail to take advantage of the work performed at their own in-house research labs. Researchers at IBM's Watson Research Center and at Xerox's Palo Alto Research Center have been phenomenally productive; they have developed some of the most innovative ideas of the information age. But these ideas are rarely translated into products.

The problem is that no one makes the connections between the technology and the market. R & D people live in their own isolated worlds. They talk to each other, attend conferences and lectures together, write papers together. They become part of an inner circle. But rarely do they spend time with customers: visiting customers, asking their opinions, inviting them to see what's going on in the research labs. On the other hand, marketing people rarely set foot inside the labs, and don't communicate with researchers very well when they do.

Successful challenger companies coordinate knowledge of technology and markets. They match technological advances with market needs—and they do so quickly and efficiently. Consider Adobe, one of the key software technology companies that helped create the desktop-publishing market. Adobe was founded by ex-Xerox PARC engineers who recognized the significant advances in computer graphics and laser printing technology. They developed a universal way of translating the graphics and text you see on your computer screen onto paper. What you see is what you get. Maybe better. Adobe was at the vortex of the market and the technology: it could see the opportunity and take advantage of it. As a result, Adobe's Postscript software has become a leading product in the new market.

Computer companies need to seize opportunities quickly. Technology is pervasive and universally accessible. If a company doesn't take advantage of its technological advances, some other company certainly will. The more time it takes to achieve product and market success, the more difficult it is to establish leadership. The rewards go to those who see the opportunities of the market and get there first.

■ **Adapt to meet customer needs.** Management consultant Tom Peters has championed a customer-oriented approach to marketing. He suggests a total dedication to "living and breathing the customer." This is good advice. But Peters means more than just listening to the customer. After "living and breathing the customer," you must be able to change and adapt your products to meet the customer's needs.

Small companies seem to do this better than large companies. Small companies tend to talk to customers at the early design stages, and they meet top management to top management. Big companies, by contrast, usually meet with customers not to exchange ideas but to win sales. Successful challengers are generally more willing to change, adapt, and alter their products in response to what customers say.

Adapting products is more of an art than a science. A vendor must make judgments about what the market wants. It must begin by developing products that are easily adaptable; then it must judge when and how to adapt them.

To do that, all parts of a company must work together as a well-coordinated unit. At many companies, each functional unit—manufacturing, development, sales, distribution—is focused on its own narrow tasks. Employees rarely see how their efforts affect the entire organization. To adapt products to meet customers' needs, a different approach is needed. All functional units must work together. In short, marketing must be everyone's job.

IMPORTANT RELATIONSHIPS

POTENTIAL CUSTOMERS
THE TRADE AND BUSINESS MEDIA
THE FINANCIAL COMMUNITY
INDUSTRY CONSULTANTS AND PUNDITS
CHANNELS OF DISTRIBUTION
ALLIANCES
THIRD-PARTY SOFTWARE COMMUNITY
CUSTOMER BETA SITE REFERENCES
YOUR EMPLOYEES AND SALES ORGANIZATION
COMPUTER COMPANY

■ **Work with third-party developers.** Vendors don't have to do all the adaptation work themselves. If a vendor builds its product right and sets up the right sorts of relationships, it can rely on third-party developers to do much of the work adapting the product to the needs of particular markets.

Consider the case of Convex, the startup that pioneered the minisupercomputer market. Convex built a superfast machine, but it didn't have the time or resources to adapt the machine to the needs of particular customers. So before it ever introduced the machine, Convex began working with independent software developers. Each software developer adapted the Convex computer to a particular vertical market. One wrote software for the mechanical design market, another for the molecular chemistry market, yet another for the geophysical market.

Convex couldn't have gone straight to the market. Customers wouldn't have bought Convex's hardware, no matter how fast it was. But Convex had the foresight to build flexibility and adaptability into its product (and its product strategy). So third-party developers found it relatively easy to adapt Convex's machine, and the machine won acceptance in a broad range of markets.

■ **Build relationships with customers.** In the past, computer customers were typically locked into a relationship with a single vendor, usually IBM. But computer users are becoming fickle. They now feel much freer to switch from one vendor to another. As customers become more computer-literate, they feel more confident buying equipment from "nonbrand" vendors. And the emergence of industry-wide standards (such as the Unix operating system) is making it easier for customers to mix and match equipment from different vendors. Customers are now able to shop around for the best buy.

In this environment, building a loyal customer base isn't easy. Companies must establish personal relationships with their customers. The computer industry is rapidly becoming a service-like industry, but computer companies have a long way to go before they become good service marketers. Service marketing is quite different from product marketing. The heart of service marketing involves personal attention, two-way communication, a focus on solving particular problems, and tangible evidence of value added. Computer makers must learn to think about marketing in a new way.

It is interesting that we tend to remain loyal for long periods of time for certain types of services, such as those provided by doctors, lawyers, and accountants. We don't change these relationships very often. The computer industry can learn a lot from these other service industries.

There are many ways to build relationships with customers. A company can set up user groups, joint marketing panels, and joint training programs. Teams from engineering and marketing can go visit the customer. R&D people should hold regular meetings with customers to get their opinions on new products. Vendors should also invite customers to workshops, joint quality programs, executive-to-executive strategy workshops, and so on.

In these meetings, information should flow in both directions. The two sides should *truly communicate,* not just send information back and forth. Building relationships is not just for show, or to make customers feel good. Both sides have a lot to gain. The customer learns about the vendor's future directions, and the vendor learns how to adapt its products to better fit the needs of the customer.

In short, vendors should make the same commitment to their customers that they make to their employees. Customers should be involved in training, development, communications—they should be viewed as full members of the corporate team.

■ **Integrate the customer into the design process.** The computer industry, through "user groups," has been a pioneer in bringing the customer into the design process. Unfortunately, communications between vendors and user groups often go in only one direction: vendors provide users with information about new designs, but users get very little input. Successful challenger companies make information flow in both directions: they integrate the user into the design process.

This trend is most notable in the semiconductor industry. Today, systems designers at computer companies play a major role in the design of new computer chips. There are at least two reasons for this trend. First, new technologies are shifting design work from people to computers, and are making it easier to unbundle the design of chips from the manufacturing process. Thus, users can more easily design their own proprietary chips: they have access to the same design tools as suppliers.

A second factor is the increasing competitiveness of computer markets. This competitiveness is pressuring computer designers to find new ways to differentiate their products in terms of functionality and performance. By designing their own chips, systems designers can more easily differentiate their products.

This same shift toward user involvement in the design process is starting to occur at the computer user level. End-user markets (such as retailing, banking, financial services, distribution, and manufacturing) are becoming increasingly competitive, causing computer users to look at their systems as strategic, proprietary weapons. In an effort to gain some control over their computer destinies, computer users are trying to gain more and more control over the computer design process. Computer vendors must respond by integrating the customer into the design process in some way. They must develop a team approach, working side by side with customers on new computer designs. As Max Hopper of American Airlines puts it, "In the future, the user will be the architect of the computer system."

I saw an example of this trend recently, while walking through the design lab at Convex Computer. I stopped to ask one of the designers about the workstation and the software program he was using. He told me Convex was a beta site for this particular system. The designer noted that the system had started out "full of bugs" and was not quite what Convex had been hoping for. But Convex designers helped the supplier reengineer the system, and now the designer described the system as "perfect." It was clear that the Convex designer felt a sense of ownership, since he and his colleagues had helped perfect the system. There is a good chance that Convex will continue to buy from the supplier for many years to come.

❏❏ Case Studies

In the rest of this chapter, I examine how three particular challenger companies put these market-driven strategies to work.

❏ *Sun:Selling Engineer to Engineer*

To keep in touch with its customers, Sun does monthly user surveys and sends senior management on sales calls. But perhaps its most effective method of staying close to customers is its policy of selling workstations engineer-to-engineer. "They're all very, very confident engineers that we're selling to," says Sun president Scott McNealy. "And so what we do is we bring them in here and we put our engineers down in front of them. It gets our prima donna engineers to know what the customer wants directly, as opposed to the sales guys or the marketing manager coming in and saying, 'You will put this work on that operating system or this feature on that piece of hardware,' or whatever. The customer tells the engineer directly."

The Sun approach changes the role of product marketing managers. At Sun, product marketers play the role of coordinators in the development process. They help package the product for the marketplace, making sure that the whole program comes together. But perhaps most important, they put the right engineers in front of the right customers. "It's more of a coordinator's role in the product marketing world, and it should be," says McNealy. "That's our view of it here. Some people may say that's a lower level view. Well, if you want to be a product marketer that goes off and invents and identifies new markets and all the rest of it, Sun isn't the place to go. You can go to a company where everybody does that and it'll drive the company right into the ground."

McNealy continues: "Bottom line, it's the customer and the engineer in this business, when you're dealing with this kind of change. If there's not a good solid link between the customer and the engineer, and if you get a bunch of people in the way of that, it's a problem. I don't care if you have lots of people up and down: when it gets down to the customer and the engineer, you'd better have a direct line there. There can be managers and product marketers and sales people and sales managers and all the rest of it around that loop, and you can go ahead and try and send stuff through that loop. But the bottom line is that we just make sure that the connection between customer and engineer happens in a very direct way."

❏ *Tandem: An Executive Institute for Customers*

Tandem has an innovative way of keeping in touch with customers. It runs a program called TEI, for Tandem Executive Institute. At a typical TEI session, Tandem will invite twenty-five or thirty customers—maybe a high-level person from one of its customers in the oil industry, another from a big retailer, another from a big bank, and so on. Tandem will also invite a professor or two from Stanford Business School to make presentations.

"It is the best marketing program we have," says Tandem president Jim Treybig. "We have some pretty aggressive discussions about point of sale or other areas of business. It's amazing. The Stanford people present case histories and set up role models where the petroleum company guy has to be a banker. They really get into it. Do you realize that there are supermarkets that handle more checks than Bank of America in a day? What is clear is that many of the businesses are converging. At TEI, our customers can really examine their own strategies, and we are sitting there listening."

Treybig also meets with customers individually. He estimates that he spends about a quarter of his time in meetings with customers. The meetings are designed to help end users define the futures of their companies, and to determine how Tandem equipment might fit into those futures. Says Treybig: "We sit there all morning and just talk with some high-level person from operations and someone high-level from MIS. And really just from talking we learn. But they learn too—because most companies don't sit and think about how ridiculous some of the things they do are. We're always talking in terms of where they are trying to be with their business. What are they trying to do?" The biggest challenge is helping customers recognize that the future can be fundamentally different from the past. "It is hard for people to go back and forget the past," says Treybig. "I mean, just forget everything they have done."

Many Tandem product strategies are based on feedback from the customers. For example, when Tandem customers began talking about the need for networking, Tandem adapted the architecture of its NonStop family of computers to allow easier networking and transactions processing. In that way, Tandem became the industry leader not only in fault-tolerant processing, but in transactions processing as well.

□ Cray: Avoiding the Growth Mentality

At many companies, growth of sales is the Number One priority. But pressure to grow can have adverse effects on a company. It is hard to chase growth as an end it itself while simultaneously listening patiently to customers, hearing their concerns, adapting to their needs, and changing the way you do business to suit the new environment.

Cray Research is steadfastly opposed to making product decisions based on sales-oriented motives. It builds its machines to order—one at a time—and lets growth take care of itself. "At Cray, we've never set a growth objective, ever," says Cray chairman John Rollwagen. "I have an aversion to that."

Rollwagen argues that most companies try to grow for the sake of growing, and they hurt themselves in the process. He faults the planning process used at most companies. The process works like this. First, planners project forward from where the company currently is. Eventually, the growth tops out and reaches some asymptote. Next, they project ahead based on a "desirable" growth rate for the company—15 percent, 25 percent, 30 percent. Inevitably there will be a difference between the two projections, between the extrapolation of current activities and the growth curve. This gap is called the *planning gap*. "The whole purpose of the planning process," says Rollwagen, "is to say, 'How are we going to fill that gap?' And my concern about filling the gap is that if I'm really going after the gap, I'm going to compromise something in order to fill it. We've got to grow more, we have to achieve more than it looks like we're going to achieve. Therefore we have to fill the gap, and how are we going to do that?"

Rollwagen says that companies often make bad decisions when they try to fill the planning gap. They start thinking things like: "Maybe we should lower the price and increase the volume and we'll have a larger company. Maybe we should diversify into some things that are related to what we do. Sure, we don't know that much about it, but we can build on what we have and diversify and therefore build a greater spectrum of products that will automatically take us into larger market opportunities—and we'll fill the gap."

That type of thinking leads to bad compromises. People are drawn in new directions not on the basis of new ideas or new technologies or new customer solutions, but simply on the basis of growth. "I can never see that process doing anything creative. I can only see it being diversionary and diluting activity," says Rollwagen. "On the other hand, I have great confidence that if you pursue good ideas as hard as you can and in a creative way and actually create new things, even though you may not know exactly what they're going to be used for, how it's going to work, they themselves will create the market and you will grow."

Rollwagen admits that his philosophy is sometimes perceived as perverse, even by his own people. But the strategy has worked for Cray: the company has grown without explicitly trying to grow. "Looking forward," says Rollwagen, "I don't measure our success by how big we are next year. I can be perfectly satisfied if next year— and I really mean this—I can be perfectly satisfied, certainly with next year's results, from a financial standpoint, if we're the same size that we are today." That is, as long as Cray's computers continue to be perceived as the highest quality products on the market. "If we can maintain the quality of our business, that's all that counts," says Rollwagen. "Size doesn't count. We can manage the company to whatever size it needs to be in order to satisfy that basic thing—what we do is make the world's fastest computers."

Rollwagen isn't worried about having to predict how many supercomputers the company will sell. "We're gonna sell precisely as many as people want to buy and no more—and, hopefully, no less. I want to sell as many as people want to buy, and I want to find all those people, wherever they are. But that's what we're gonna sell and there's nothing we can do about that; that's Mother Nature. So we can accept that. I want all the growth that's there, but there is only so much. And that's fine. To arbitrarily say that this company has to grow at x percent a year in order to be successful is really dumb, in my opinion."

❏❏ A New Type of Customer Loyalty

In the diverse and changing world of computers, there is always going to be someone who will provide a specific solution, who will offer the customer a better way. In bygone days, computer users were tied to particular computer systems because of their investments in particular operating systems. But users don't want to be tied down anymore; they too live in a changing, highly competitive world. They must select computers that enhance their productivity, make them more cost-effective, or give them a service capability that their competitors don't have.

So computer companies must look to new means of maintaining customer loyalty. They must listen to customers and respond to their needs, providing creative solutions and adapting their products to fit the changing computer environment. Being market-driven is a matter of listening, observing, and responding to the customer. Computer companies must understand the real, competitive needs of their customers and take the responsibility for providing solutions.

You have to know who to look in the eye.

Frank Perdue

□ 8

Relationships

■■ THE COMPUTER BUSINESS is ultimately a battle of resources. But in today's environment, that doesn't mean that the company with the most money wins. Instead, it means that companies must focus and invest their money in resources that provide maximum return. For most companies, and for small companies in particular, strategic relationships have become the best way to maximize return—and to survive in competitive markets. Indeed, most of today's successful computer products result from the coordinated efforts of several companies. Consider the following:

■ The Apple LaserWriter combines a laser engine from Canon, software from Adobe, and product design and marketing from Apple. By working together, the companies brought the graphically oriented laser printer to the market long before any competitor, helping Apple launch a wildly successful new market: desktop publishing.

■ Tandem is the leader in fail-safe computer systems. But the company has little expertise in the area of telecommunications. So Tandem made an investment in Integrated Technology Inc., a small telecommunications company. Together, the two companies have developed products to link together Tandem's NonStop computers.

These are examples of *strategic relationships*. In each case, companies are working together to create products that none of them could have done alone. These types of relationships are increasingly important in today's computer industry. Even IBM is recognizing that it can no longer do everything itself; it must link up with other companies.

In this chapter, I discuss the reasons that strategic relationships are becoming increasingly important for high-technology companies. Then I offer suggestions on how to establish and build successful relationships—including a look at why relationships with IBM often prove dangerous.

❏ ❏ The Need for Relationships

For many years, vertical integration was the name of the game in the computer industry. Companies tried to do everything themselves. They made the electronic components. They produced peripherals like printers and disk drives. They developed the systems software. They sold the finished systems. Everyone agreed: vertical integration was the way to build and maintain a successful computer company.

There are several advantages to vertical integration. Within vertically integrated companies, designers of components can collaborate with the end users of their products, namely the systems designers. This collaboration can lead to designs that are specialized for high performance in particular applications. Vertically integrated companies can also use their internal markets to establish scale economies on certain components or subsystems, bringing down the prices of products that they sell to outside customers.

In today's computer market, however, there are new factors weighing against vertical integration. Increasing costs of developing and bringing products to market, and rapid changes in products and technology, are making it more difficult for any company to develop everything by itself. Computer companies must get products to the market more quickly than ever before—and the products themselves are more complex than ever before. Minicomputers, for example, started out as simple circuit boards for dedicated technical applications. Today, minicomputers are powerful enough to run entire factories and coordinate the worldwide operations of major corporations.

Even personal computers are incredibly complex machines these days. A decade ago, personal computers were simple to design and build. But in the drive to make personal computers more useful, companies have developed sophisticated operating systems and applications software, larger and more reliable information storage products, and more efficient ways of putting information in and getting it out. In addition, companies have developed new networking schemes to allow personal computers to share information within the office, in the plant, or throughout the world.

Customers now expect all these features when they buy a computer. But with products and technologies changing so rapidly, and with dozens of small companies focusing on every narrow aspect of the technology, no one company can possibly stay at the forefront in all areas.

As a result, vertical integration may be going out of style. Vertically integrated companies tend to lose touch with leading-edge technologies and developments. Studies have shown that small companies were responsible for about 55 percent of the recent innovations in the United States. Thus, if a company tries to do everything on its own, it will likely miss the next wave of technology, and it will get stuck playing catch-up forever. Companies cannot afford a "not invented here" mentality. They must buy and borrow leading-edge products and technologies that are developed elsewhere.

Japanese companies have addressed this problem by buying technologies from U.S. companies. According to one estimate, Japanese companies acquired 32,000 technology licensing agreements, mostly from the United States, between 1950 and 1978. It is estimated that the Japanese paid $9 billion for these licenses, but it gave them access to technologies that cost the developers more than $500 billion. While acquiring these product-related technologies, the Japanese targeted their own research efforts on process and manufacturing technology. This strategy has won the Japanese ownership of many electronics markets.

For most companies, though, strategic partnerships make much more sense than outright acquisition of technologies (or outright acquisition of other companies). Among challenger companies, strategic partnering takes place very frequently and very openly. Companies form strategic relationships for a variety of reasons. They might want to:

■ Acquire components or software for their products.
■ Broaden their financial base.
■ Create a new position.
■ Expand their technological expertise.
■ Increase their contacts within the industry.
■ Extend their products into new application areas.
■ Expand the distribution of their products.

Whatever their reasons, companies can expand their opportunities considerably when they join together. Relationships can take many different forms. Companies can exchange various assortments of technologies, products, money, and stock. The only ground rules are that both companies get something out of the relationship that they couldn't have done alone.

The old BUNCH mainframe companies, for example, have transformed themselves by acquiring leading-edge products from other (usually smaller) companies. NCR, Honeywell, and Unisys (the merged combination of Burroughs and Sperry-Univac) have all developed strategic relationships. In most cases, these companies buy completed products from small companies, integrate the products into their own product lines, and resell the products into their existing customer bases.

Unisys, for example, uses Convergent Technology's workstations in its own B-20 series of computers. It has also announced agreements with Sun and AT&T to gain access to the Unix community. "All of our new high-growth markets are supported by strategic alliances," says senior vice-president Jan Lindelow. "We have historically provided expertise in systems development and vertical market software, but without our relationships with other companies we might have missed important market opportunities."

Honeywell-Bull, meanwhile, is the result of a "superalliance" combining France's Compagnie de Machines Bull, Honeywell's computer operations, and Japan's NEC. Together, the three companies have a combined R&D second only to IBM's. And they have market connections in Europe, the United States, and Japan. "We may be the world's largest startup," says Jerry Meyers, president of the joint venture.

Strategic relationships are particularly important for small companies, since their resources are particularly limited. It makes sense for innovative young companies to focus on what they do best—and to acquire the rest from other companies. There are about 10,000 companies developing products for the computer market. Those companies cover every market segment, every peripheral, every technology. No small company can compete across the board in all market segments, nor can it stay at the forefront in all technological areas. By acquiring products and technologies from some of the other

10,000 computer companies, new companies can get their own products to market faster—and postpone their "make or buy" decisions until later, after they have established market positions.

Digital provided an early example of the power of relationships with the network of original equipment manufacturers (OEMs) that it developed in the 1970s. The OEMs took bare-bones Digital machines and customized them for particular applications and particular market segments. Digital worked out relationships with hundreds of OEMs, each of which extended Digital's reach in the marketplace.

Dave Martin, president of National Advanced Systems, notes that the climate for relationships has changed drastically since the 1970s, when people felt "the only place to do things is inside my company." Now, says Martin, people use the following reasoning: "I know what I want to do, I know what the customer wants, and I've got to fulfill it. How do I do it in the most pragmatic fashion to keep my own internal resources properly balanced with external partnerships?"

Martin says that his company did about a dozen partnerships during 1987—and it turned down many other opportunities. Martin estimates that he gets ten calls a week about partnership opportunities. "Other than the Hitachi partnership, we had nothing up until a year ago," says Martin. "In the past eleven months, we had twelve, all of which fit very nicely and which I really do believe, with maybe one exception, are mutually advantageous."

Martin describes a typical partnership like this: "We get your networking technology and we give you vector-processing technology. Of our twelve partnerships, ten have been that type. Both partners are thinking that they could get something that would make their situation better off than if they tried to do it themselves."

Martin finds the concept of partnerships so important, in fact, that he created a new office at NAS called Strategic Plans, Programs, and Partnerships. He says he needs the new function partly because there are more computer companies in the business than ever. "A lot of them aren't fully integrated companies—you know, they don't have full development, manufacturing, sales and service, and worldwide distribution—but a lot of them have pockets of product or distribution capability. So, I think you can out-partner IBM. And I think partnering is a very pragmatic and practical way to go about things. If anything, I think it'll become even more so."

❏❏ **Relationships with IBM**

Throughout most of its history, IBM tried to do everything itself. It designed and manufactured all of the components for its computers, and it controlled the distribution of the machines. IBM's culture has strongly discouraged technology sharing. It has been a closed corporation, operating as if it were self-sufficient. To a large extent, this strategy worked: for many years, IBM was able to draw on its own resources for whatever it needed to dominate the computer industry.

As the computer market has changed, however, even IBM has reevaluated its thinking on external relationships. As computers have become more complex, and technologies have changed more rapidly, IBM has recognized that it cannot do everything itself. Now IBM has a new approach. It looks inside first, but when it finds markets that are strategically important but technologically out of its grasp, it buys technologies, products, or even companies to gain access to those markets. IBM's original personal computer provides an example. As I discussed in Chapter 3, practically the only thing that IBM added to the PC was its logo.

In some cases, relationships with IBM can work out well for both sides. Intel and Microsoft have clearly benefited from their relationships with IBM. Indeed, the IBM "seal of approval" helped turn Intel and Microsoft into industry leaders.

But other companies have been burned by their relationships with IBM. Unlike other technology companies, IBM enters its relationships from a position of power and control. In some cases, it has used partnerships to learn new technologies and then has abandoned its partners, leaving them in a weakened condition, to fend for themselves. More than a few companies have entered relationships with IBM full of hope and pride, but have ended up frustrated and angry—and financially troubled.

Gene Amdahl argues that IBM's relationships have, in many cases, benefited only IBM. Amdahl recalls one instance involving the semiconductor industry. Until the late 1970s, IBM had produced all the semiconductor chips it needed for its computers. But then, says Amdahl, "IBM assessed their situation in the semiconductor world and found out that they were well behind in almost every area. So they let out a series of contracts for R&D." IBM also began purchasing chips from leading semiconductor companies, and used the chips in several new low-end computers in its 370 line. IBM's demand for chips was so high that other computer manufacturers had trouble getting chips from the semiconductor makers. "People who didn't have their memory chip orders in to the semiconductor companies found that they couldn't get in," recalls Amdahl.

Up to that point, this was good news for the semiconductor companies. IBM was a huge new customer, and the semiconductor companies looked forward to a long relationship. But it was not to be. Through its R&D contracts, IBM gained the technological expertise that it needed, so it shifted its chip production back in-house. "It was a big letdown for the semiconductor industry," says Amdahl. "They had taken the bait very nicely, and they went into recession."

There was a similar incident five years later, when IBM began producing its personal computer. For a while, IBM purchased many of its disk drives from MiniScribe, a small Colorado company. At first, the deal was great for MiniScribe: it became a leader in the disk-drive industry, and its stock soared. But then IBM decided to buy its disk drives from somebody else, dropping MiniScribe entirely. MiniScribe's stock plummeted by more than one-third. Just as many electronics companies have gotten into trouble by relying too heavily on the Pentagon, they can also get into trouble by relying on Big Blue.

Some IBM challengers credit IBM with using partnerships effectively. "I would say that in the early '80s, IBM was a totally remade company," says Dave Martin of National Advanced Systems. "They were doing partnering better than anybody." But Martin would much rather form relationships with other small companies than with IBM itself. "IBM is basically trying to create a bigger IBM by using money. So it's different than smaller companies' partnerships," explains Martin. "If you get hooked up with IBM, is it good news or bad news? There are so many skeletons along the road now."

❏❏ Case Study: Relationships at Apple

Relationships among challenger companies tend to be much more balanced than relationships with IBM. Apple, particularly under the leadership of John Sculley, has been a pioneer in the development of strategic relationships. According to Sculley, no company, no matter how powerful, no matter how large its customer base, can succeed without a network of relationships. "The network is far more inclusive than just your own company. The network includes all the people who you have relationships and dependencies with around you," says Sculley.

Although Apple is in a strong position in the industry, Sculley has no expectations for it to be totally self-sufficient. "The old paradigm was that you had as much self-sufficiency as possible," says Sculley. "Even the CEO was expected to be invincible. He was the John Wayne hero model from World War II that we would follow. Today, we learn that CEOs are vulnerable. They make mistakes. We expect it."

Rather than having Apple do everything itself, Sculley has tried to set up relationships with other companies that are very good at doing particular things. "When you do everything yourself, in the short term you may get better margins, but you also lose tremendous flexibility to be able to change," says Sculley. "And as hard as we work to try to define what the information technology industry might look like by the beginning of the next century, we still can't do it with much accuracy. We want to retain that flexibility of being able to change as circumstances change."

Apple began forming strategic relationships in the early days of the Apple II computer. For example, Apple's team of "education evangelists" worked closely with third-party software companies that wanted to design educational software for the computer. As a result, Apple gained a dominant market share of the educational market. More recently, Apple has joined forces with Adobe and other software companies to help launch the desktop-publishing market.

The boldest, and perhaps most important, of Apple's relationships is its new co-development alliance with Digital Equipment. In January 1988, the two companies agreed to work together to develop products to allow Apple's Macintosh computers to connect and communicate with Digital's VAX minicomputers. Together, the two companies hope to compete more effectively with IBM in the fast-growing market for distributed computing.

Increasingly, users want to be able to connect personal computers to larger machines, using the smaller computers to retrieve and manipulate data that is stored on the larger machines. The Macintosh-VAX linkup will make that possible. Customers will be able to fill all of their distributed computing needs from Apple and Digital, without ever buying a thing from IBM.

By joining forces, Apple and Digital realize several benefits:

▪ **Bigger markets.** The link between the Macintosh and the VAX will open up new applications for both machines. Each will be used in situations where it would not have been used otherwise.

▪ **Multiple sales organizations.** Apple's sales force will promote networking with the VAX, and Digital's sales force will do the same for the Macintosh. By building connections to Digital, Apple gets an effective sales force that is hundreds of times larger than its actual sales force.

▪ **Greater user confidence.** People will be more confident buying from Apple and Digital, knowing that they can expand to a complete network solution if they desire. Indeed, the day after the Apple–Digital joint announcement, sales of Macintosh computers rose sharply—a clear expression of user confidence in the future of the Mac.

Apple's relationship with Digital is a rather typical strategic alliance. But Apple has also pioneered a totally new type of strategic alliance: a relationship with a spinoff company. In 1987, Apple took all of its application software groups and spun them off into a separate and soon to be independent company called Claris. The spinoff is headed by Bill Campbell, former executive vice-president of sales and marketing at Apple. Claris is now responsible for the enhancement and marketing of such Macintosh hits as MacWrite, MacDraw, and MacPaint. Campbell has assembled an impressive array of software talent to develop new proprietary products of its own.

Actually, Sculley resists the term *spinoff*. He insists that Claris doesn't fit the traditional model of a spinoff company. "In the industrial age model, things that didn't fit were spun off," explains Sculley. "In the information age model, we will conscientiously *spin out* but not *spin off* businesses. Claris is a good example. That's not a spinoff. It's not 'goodbye and good luck.' Rather, we're really saying that there is a piece out there that we would like to have in the network, we don't necessarily have to own it all, but we want it to be very successful. And we'd like it to have some characteristics which will be oriented toward the things that are important to us. So we will give it our genetic code by taking a management team from Apple, we'll give it some of our resources, we'll give it some initial critical mass to get started and spin it out, but we expect it to be a very active participant in that network. If that model works, I think between now and the end of the century that you will see a number of other examples of our creating companies and spinning them out."

The key question that companies must ask themselves is: "What are we good at?" Companies should form relationships based on the answer to that question. In Apple's case, says Sculley, "We have demonstrated that we are good at creating a company, but we haven't demonstrated yet that we are as good as IBM at managing a company. So, in leveraging our strengths, why not continue to do what we really do well, which is to create companies? The spin out, as opposed to the spinoff, is one model which we are going to follow."

❏❏ Taking the Plunge—Carefully

Although strategic relationships are vital to success in today's market, they are by no means a guarantee of success. Indeed, a poorly conceived relationship can cause far more damage than good. Thus, it is important to look (and think and plan) before you leap into relationships—and to continue to support the relationship after it is formed. Otherwise, the relationship will be a waste of time and resources.

By following a few simple guidelines, companies can greatly increase the chances for success in their relationships.

■ **Know your strengths and weaknesses.** Companies should enter relationships to enhance their strengths, not to correct their weaknesses. If you enter an alliance from a position of weakness, your partner will have all the leverage, and you are likely to end up on the short end of the relationship.

Instead, just as in a good marriage, both sides should have strengths. The relationship between Apple and Adobe is a good model. Both companies entered the relationship with real strengths: Apple with its Macintosh interface technology, Adobe with its Postscript software technology. Together, they were able to build a more powerful desktop-publishing solution for customers.

■ **Establish a clear set of responsibilities.** Many relationships fail because not enough operational details are worked out ahead of time. Once the relationship is underway, the two sides end up arguing over things like how many engineers each side will contribute to the joint effort. Before signing any agreements, partners should decide who is going to do what, and how it is going to get done.

■ **Find out more about your partner.** If you are about to strike a deal with a new partner, talk to other companies that have dealt with them. Do they live up to agreements? Do they share the load? Do they frighten easily when problems arise?

■ **Involve the entire organization.** A few years ago, the presidents of Apple and Cullinet got together and formed a relationship. Cullinet, which had previously designed software only for IBM computers, agreed to revise some of its software so that it would run on a Macintosh. But the work never got done. Why? Because the engineering people from the two organizations felt left out of the deal. The top managers inked the deal and dumped the project on the engineers, without any clear set of goals and objectives. The engineers had no sense of ownership of the project, and no clear direction, so they let it languish.

■ **Keep working at the relationship *after* you take the plunge.** To make sure the relationship doesn't falter after it is formed, each side should appoint an "alliance champion" to oversee the project on an ongoing basis. In addition, senior managers from the two sides should meet regularly to review the progress of the alliance. Relationships are a tricky business. Each side has its own goals and objectives. Without constant attention, the alliance is likely to get off track.

The Network is the System.
Scott McNealy, Sun Microsytems

□ 9

Standards

■ ■ A FUNNY THING is happening to standards in the computer industry. They keep changing. It's one of those paradoxical situations: standards seem to be creating more diversity. While old computer standards restricted customer options, new "open standards" expand the possibilities for computer users. With the new standards, computer users are no longer restricted to a single brand of computer; they can mix and match computers, and still share information among all of their machines.

When the computer industry was primarily a mainframe business, the words standard and IBM were pretty much synonymous. Other computer companies waited for IBM to announce new standards. Then, after new standards had been set by IBM, competitors tried to follow quickly with products of their own. Companies who offered new technologies without IBM's blessing were met with skepticism in the marketplace. Standards were established to keep competitors out, or at least to keep them far behind the standard bearer.

What we call standards today are quite different from standards of the past. In today's computing world, a standard is a generic entity that can be shared by many different companies. Consider, for example, the Unix operating system, one of the most important new standards. Developed by AT&T's Bell Laboratories, Unix has enabled user companies to declare independence from computer solutions confined to single product lines. Designed to be hardware-independent, Unix can run on any suitably powerful computer. As a result, computers of all types and brands can communicate when they run the Unix operating system. With Unix, users can design unique networks tailored to their own requirements, sharing resources like documents, files and printers.

Standards today allow computers of all types to talk to each other. A user can connect a Cray supercomputer, a VAX minicomputer, a Convex minisupercomputer, a SUN workstation, and an Apple

Macintosh. Instead of excluding competitors as IBM's standards once did, the new standards encourage differentiation and the proliferation of individualized computer networks. Today's standards create vast opportunities for computer vendors offering solutions for specific tasks. These vendors can chip away at the customer base of IBM, which is still living in the past, offering top-to-bottom solutions using only IBM products.

New technologies no longer need IBM's validation to survive in today's computing environment. In many cases, in fact, IBM has become the follower, initially reluctant to adopt new technologies that might damage its monolithic market share, but eventually adopting them anyway in a never-ending effort to catch up and stay competitive.

❏❏ The Need for Open Standards

Computer standards have evolved over the years, driven by the radical changes in the computer marketplace. At one time, the computer market was very orderly: there were mainframe computers and not much else. Almost everyone bought mainframes from IBM, so the IBM standard reigned supreme. But as the computer market began to fragment, with hundreds of companies offering specialized computing solutions for niche markets, the market moved towards chaos. Dozens of new "ministandards" emerged, none of them compatible with the others.

The changes began as companies chipped away segments of the traditional mainframe market. Digital's first minicomputers, sold at a fraction of the price of IBM's mainframes, distributed computer power to engineering and technical professionals. Personal computers, priced still lower, dispersed access and control to a massive base of general-purpose computer users. New companies and products entered the market every day, offering computer users a plethora of new options. As computer users gained experience with the new machines, they increasingly based their buying decisions on functionality, not brand name. For each new application, they surveyed the market and bought the computer best suited to that particular task.

The result was a crazy quilt of computer solutions. Individual companies bought dozens of different types of computers, none of which could share information or peripherals. An article in the August 1986 issue of *Barron's* aptly describes the situation:

> *On the typical U.S. factory floor the problem can be truly daunting: A succession of plant managers bought automated equipment as it became available, from whatever vendor invented it or made the best sales pitch or offered the best price. The result: "islands of automation" in which parts are machined or assembled efficiently and then cast adrift. Things are certainly no better in the office, where word processors, accounting systems, sales and marketing and inventory controls systems frequently work at cross purposes or not at all.*

The rapid pace of technological change has made the problem even worse. As soon as a company buys a new computer system from one vendor, another vendor typically introduces a new machine, even better suited to the customer's needs. But the customer is out of luck. Because most existing computers are closed systems, incompatible with the rest of the computing world, users cannot connect them to newer computers. Thus, the customer is stuck with obsolete equipment, with no easy way of upgrading.

To drive home the absurdity of computer incompatibilities, Sun chief executive Scott McNealy compares computers to automobiles. "Can you imagine going and renting a car in Paris, and you had to read a stack of manuals three feet high and you had to have a master's degree in automobile operation? Imagine you had to go to school for six years at a place like MIT, and you had to read a manual for, say, three months, and you had to memorize that manual—and then you could be 98 percent productive on that Paris automobile. That's what it's like to switch from one computer to another." With a car, says McNealy, the situation is totally different: "When you sit and try the car, you're 98 percent productive on that automobile within ten seconds without reading anything. You can just sit down and you know what you do. You look around, you scan it, and you're ready to go."

McNealy uses a similar analogy to highlight the limitations of today's computer networks. "Imagine that you landed in San Francisco and you wanted to drive to Apple's headquarters—but you needed to drive a Ford because only Fords had a structure that would go on the network of roads to Apple. But then if you wanted to come over to Sun, you had to go re-rent another car, a Chevy, because only Chevies could go on the road to Sun. That's the networking environment that we were dealing with in the early '80s."

McNealy continues the comparison: "The other question I ask people is, 'I just invented a new car. It's available to you for $200. It goes 200 miles an hour, gets 200 miles to the gallon. Would you like it? Would you buy it?' Most people say, 'Of course!' Then I add, 'I have to tell you it's fifty feet wide.' Then they say, 'Oh, I don't want one of those. I'd need a runway!' That's the kind of thing that all the new computer startups were coming out with. They'd say: 'I've got this new 50-foot-wide car: have at it, go for it, it's a great product.' But nobody could use it. Or if they used it, they had to go out and build runways. And then, if the people who were building the cars went out of business, the users had invested all that money in runways and now they were just wasted runways."

McNealy's central point: incompatible systems lead to all sorts of inefficiencies. Among computer users, incompatible systems contributed to significant losses in productivity. Functions within the corporation became increasingly disjointed, even as individual groups became better at their particular tasks. Individual workers sat at their personal computers doing specific tasks. If they wanted to share information with people in other parts of the corporation, they had to go through all types of contortions, such as physically swapping floppy disks, or even retyping data.

To improve productivity, computer manufacturers must increase computer speed, simplify computer use, and facilitate communication among computers. Computer speed has been improving steadily, but ease of use and communications have lagged behind. Companies need to network their computers together, so that people on the factory floor, in the production facility, and in retail distribution can share information. Such connections can make workers more efficient and more responsive to customers. In today's society, where product life cycles are shorter than ever before and competition more intense, time is always at a premium. Today's emphasis on computer networking flows directly from the drive to improve productivity through efficient communication. Networking mainframes, minis, and PCs, or linking the hierarchy of supercomputers, minisupers, and workstations, are all efforts to improve communication and leverage investment in computer systems.

Another force driving the networking trend is the changing structure of corporations. In old-line corporations, everything had its place, and tasks were passed off from one functional organization to another. A corporation seeking success in the new age of Darwinian competition must act more as a unified entity. Coordination between engineering, product development, and manufacturing is essential for getting products to market in a timely fashion. Effective marketing requires coordination between marketing, sales, and manufacturing groups. And to meet the needs of rapidly changing markets, distribution organizations must be flexible, yet efficient.

Computer networks can unify all of these corporate functions. In less time than it takes to schedule a meeting, computer networks can provide corporate decision makers with up-to-date order rates, inventory, manufacturing schedules, distribution schedules, market trends, and competitive analyses. Information from around the company can be unified to provide a more consistent picture of the company's current position. Easy access to this information is essential: delay can mean loss of competitive position, loss of sales, and loss of profits.

It's not all here yet, but strategic thinkers in corporations around the world know where they have to be in the next decade. They know that the corporation of the 21st century will have to be a fast-acting, tight organization acting as a unified force. Companies must make sure that employees have access to the information they need. They must build teamwork based on common information; they must get their employees all moving in the same direction.

Companies need to make information flow part of their central strategies. Improved information flow means improved productivity—and improved corporate performance. As Tandem's Jim Treybig puts it: "We're moving into the information age. The information age means your computer system, your information delivery system, is your competitive weapon." Customers don't want any one company to own connectivity and compatibility standards. They don't want to be restricted in their efforts to improve information flow and productivity. They need free and easy access to all types of machines as they adapt to technological changes and adopt new solutions to ever-evolving problems.

❏❏ Users Take the Offensive

Computer users, recognizing the need for improved information flow, are faced with a dilemma. They don't want to buy from a single computer vendor: no one vendor can meet all of their needs. But linking together machines from different vendors is difficult, if not impossible. They know they can't rely on computer vendors for a solution to this dilemma; without pressure from users, vendors have produced a mess of incompatible systems.

So computer users are taking matters into their own hands. They are banding together to force computer vendors to adopt *open standards*—standards that will allow machines of all sorts to work together and share information.

Computer standards can take many forms. There are standards that let computers run the same programs, standards that let computers display the same types of graphics, standards that let computers send data back and forth from one to another. Customers are pushing for all of these standards—and more.

General Motors, for instance, got tired of waiting for IBM and other computer vendors to come up with connectivity solutions for computerized factories. So GM designed its own standard called the manufacturing automation protocol (MAP). When the auto giant informed its vendors that it would not buy any new equipment from them unless the equipment was MAP-compatible, the vendors had little choice but to agree. After the U.S. government, General Motors is the largest purchaser of computer equipment in the world. The fear that other large accounts would take compatibility matters into their own hands spurred swift action by vendors. Now dozens of vendors are producing MAP-compatible equipment. The MAP standard will free GM (and other manufacturers) to shop for the best equipment possible, regardless of brand, with the assurance that the machines will be able to communicate with all other machines in the factory.

THE ELEMENTS OF AN OPEN SYSTEM

Once they're perfected, most likely in the early 1990s, open systems will consist of various pieces of software written to run on all brands of computers and to tie them into networks. The idea is to liberate computer buyers from dependency on a single supplier. Below are three key elements of open systems and the status of each

Projects	Suppliers	Availability
OPERATING SYSTEMS Manages the flow of data among a computer's components and schedules their activities.		
Unified Unix	AT&T/Sun Microsystems	Early 1989
Posix	IEEE/National Bureau of Standards	3Q '88
USER INTERFACE Interacts with computer operator, accepting input from mouse and/or keyboard and displaying results on a screen		
X Windows	MIT/X Consortium	3Q '88
Presentation Manager	Microsoft/IBM	4Q '88
Open Look	AT&T/Sun Microsystems	1989
NETWORKING PROTOCOLS 'Rules of the road' that regulate the flow of data among different computers on a network		
Open Systems Interconnection	ISO*	Partly completed
X.400 (electronic mail)	CCITT**	Exists
FTAM (file transfer)	ISO	Exists
Network File System	Sun Microsystems	Exists
Network Computing System	Apollo Computer	Exists

*International Standards Organization   **International Telephone & Telegraph Consultative Committee
DATA: X/OPEN, BW

Reprinted from May 23, 1988 issue of *Business Week* by special permission, copyright © 1988 by McGraw-Hill, Inc.

The growing user demand for interoperability and compatibility has forced computer vendors to create new alliances aimed at setting standards. One such group, called X/Open, is dedicated to establishing and supporting a multivendor applications environment, so that application programs could be run on all different brands of computer. The nonprofit consortium has thirteen members, including AT&T, Digital, Hewlett Packard, Olivetti, Siemens, and Sun. Together, the members account for 25 percent of worldwide computing revenues. The group has a user council, composed of senior executives from sixteen major computer users.

Another major alliance is the Corporation for Open Systems (COS), a research and development consortium dedicated to enabling diverse products to communicate with one another. Formed in January 1986 by the chief executive officers of seventeen competing computer companies, COS is lobbying U.S. computer vendors to support international standards such as the International Standards Organization's Open System Interconnect model and the Integrated Services Digital Network. It is also establishing a single, consistent set of tests and certification methods.

COS member companies recognize that linking their machines to the machines of their competitors will loosen the stranglehold they have on existing customers. Installed bases once claimed by single companies will be reopened to competition. But COS members have decided that the benefits of such connectivity outweigh the disadvantages. COS members are betting they can get more business in the new open market than in a proprietary one. They plan to differentiate themselves not through proprietary networks, but through unique solutions to customer problems, through integration, and through support services. They believe that the companies and computers that offer the best *solutions* will profit the most in the new open-systems environment.

Because COS had been perceived as a collective competitive weapon against Big Blue, many in the industry were shocked when IBM joined COS in February 1986, a month after the founding of the new group. But IBM's participation was actually crucial to COS's success. Big Blue controls nearly two-thirds of the domestic installed base and three-quarters of the mainframe business. Without its cooperation, COS standards could hardly become universal. For its part, IBM has been working on making its Systems Network Architecture compatible with international communications standards. By joining COS, it buys influence on how and when new protocols are developed.

Customers will continue to demand connectivity until it becomes a nonissue, until companies routinely provide equipment that can be linked in a limitless number of combinations. CCI's John Cunningham describes the kind of freedom he thinks connectivity eventually will provide companies. "There will be an information services executive that will have veto power on what devices are put into that network. A company's main information problem is basically control of the network. Therefore, there have to be some standards defining what gets screwed onto the network." The MIS executive, says Cunningham, will let people buy whatever equipment they want, as long as the equipment is able to download applications and send information along the network. For example, the MIS manager will tell people something like this: "You can buy anything you want, but you better make sure it's MS-DOS-compatible so I can download applications. And you better make sure it supports the X.400 communication protocol because that is going to be the corporate network and I want to use you as an end node on my mail network."

Amdahl's Zemke sums up his view of the future of standards. "The fact of the matter is that open standards are coming," says Zemke. "It's not a question of if, it's just a question of when. To a large degree, it will be dictated by the users: when the users stand up and say, 'Look, we're behind the Corporation for Open Systems.' Things like this MAP protocol that came out of GM—those are big users who just put their foot down, big users, and said, 'Look, we're not going to buy any more if you guys don't give us a standard that we can deal with.' And I always tell customers: 'You know, you write the check. Don't get confused about your ability to impact the industry and its direction.' "

❑❑ Standards, IBM Style

IBM has established many industry standards and published the specs for others to use, such as Systems Network Architecture (SNA), Structured Query Language (SQL), and Token Ring. And in recent years IBM has joined most of the industry-standard promoting organizations, surprising industry observers, many of whom questioned their motives. But the prospect of universal connectivity and universal compatibility threatens IBM's conventional approach to computing. To understand IBM's position, it is useful to distinguish between different types of computer compatibility.

- *Intraproduct compatibility:* Compatibility within a particular product line from one company
- *Intracompany compatibility:* Compatibility between a company's different product lines

▪ *Intercompany compatibility:* Compatibility between computers from different companies

Computer customers want the third level of compatibility: they want to use a mixture of computers from different vendors. "The third level is the level that the customer ideally wants," says Dave Martin of National Advanced Systems. "They want total interoperability of systems, so you can take heterogeneous environments, a DEC environment, an IBM environment, an Apple environment, and at the highest level the customer views total transparency."

But IBM is struggling to deliver the second level of compatibility—that is, compatibility between its own product lines. They have announced SAA, Systems Application Architecture, intended to provide compatibility from its PS/2 personal computers to its minis and its mainframes. But certain IBM products are excluded; and some believe that all the ballyhoo about SAA is the best FUD factor IBM has going for it. SAA is to be delivered on a multiyear schedule, which has yet to be published. It's difficult for IBM to provide the second level of compatibility, considering its past successful product lines and its huge intstalled base. "That's where they have had the most control, but today, it's the least attractive to the customer," says NAS's Martin.

By contrast, Digital has made intracompany compatibility a major selling point. Indeed, much of Digital's success in recent years can be attributed to the fact that *all* of its computers can work together in networks. "IBM has operated at the box level, the intraproduct level," says Martin. "They have different [incompatible] architectures [in different product lines]. DEC ate IBM's lunch because they operated up at the intra-company level."

Why does IBM operate at that low level of compatibility? In part because it *wants* to lock customers into particular architectures, to make them hostage to Big Blue. But there are other factors too. In some ways, IBM's lack of intracompany compatibility is a reflection of the company's organization and style. Although IBM's corporate culture stresses uniformity, the company paradoxically gives individual researchers a considerable measure of autonomy. By allowing these "wild ducks" to fly out of formation, IBM sacrifices uniformity in its product lines.

There is a constant tension between creativity and uniformity at any large company. Digital founder Ken Olsen speaks from experience. "We learned that if you let an organization go its own way, you end up with many different kinds of computers, many different kinds of software systems, and many different ways of roping them together," says Olsen. "With its 360, IBM had a magnificent plan. It had only one architecture, only one way of doing everything. But then they dropped that beautiful part of their plan."

Olsen continues: "Richard DeLamarter, the lawyer from the Justice Department case who wrote the book (*Big Blue: IBM's Use and Abuse of Power*) on IBM, says IBM went off in all directions in order to confuse the competition, which might be partly true, but I doubt it. I suspect that it's just the natural way of the organization. Everybody comes up with a new project, they all want to do it differently, their own way, and every department in the company, every group in the company, feels the rest of the company is their first competitor. Anybody outside is secondary."

Olsen notes that Digital had similar problems in its early days. "Every time we made a new PDP-8 or PDP-11, people made improvements that made it completely incompatible with the previous one," says Olsen. "And everybody—probably a dozen or two people—wanted a different operating system for the PDP-11. I don't know how many we had: five, six, or eight. I'm not even sure how many we turned off."

In the 1970s, these internal incompatibilities caused such confusion that Digital executives made a fundamental decision: the company would limit itself to one computer architecture on software systems. "A magnificent, obvious idea," says Olsen. "Anybody would have thought of it, except that that kind of thinking is just not natural for a large organization."

IBM made a move toward open standards with its personal computer. By using off-the-shelf chips from Intel, and an off-the-shelf operating system (MS-DOS) from Microsoft, IBM got into business quickly. Within a couple of years of entering the market, Big Blue was dominating the Fortune 1000 users of PCs.

But there was also bad news for Big Blue. The openness of the PC made it easy for other companies to make compatible computers. For customers, this was good news. The plug-compatible business provided a lot of alternatives in pricing and performance. But from

IBM's perspective, the growth of the plug-compatible business meant a loss in its treasured control over the user. The move to open architecture created a tough set of competitors that nibbled away at IBM—at a time when Big Blue really needed the revenue. Control of the market shifted away from IBM. In fact, the MS-DOS standard, which started as IBM's standard, now seems like everyone's standard. In 1986, Compaq announced its next-generation MS-DOS computer, long before IBM announced its future intentions. Five or ten years earlier, such a move would have been viewed as market suicide. Compaq's unprecedented move led industry pundits to ask: "Just whose standard is MS-DOS anyway?"

Indeed, the IBM personal computer is more Intel's standard and Microsoft's standard than it is IBM's standard. IBM helped create the standard by selecting Intel's microprocessors and Microsoft's operating system. But now anyone can use the same ingredients and build a compatible machine. Intel and Microsoft are the chief beneficiaries.

IBM recognizes that it is losing control of its customers. The company is trying to regain some control with its Personal System 2 series of personal computers. Like IBM's original personal computers, the PS/2 line is based on industry standard parts. But the machines have some added twists—such as the so-called Micro Channel communications bus—that seem designed to shut out the clone-makers.

"IBM's strategy with the PS/2 is a fascinating one because it's a reversal from the open architecture that marked their initial entry into this market," says Amdahl's Zemke. "They're openly saying that they're delivering a system that will be tougher and tougher to clone. It's fascinating because if I were a big user of those systems, it would be like someone coming to me and saying, 'I've got this set of advantages for you, but you're going have to give up freedom and competition to get it.' I wonder if people are going to do that."

Zemke isn't the only skeptic. A few years ago, customers and competitors would have fallen in line as soon as IBM announced a new standard. But that's not happening this time. As I discussed in Chapter 6, customers and competitors are questioning the advantages of IBM's new Micro Channel architecture. These days, it seems that IBM just can't win, whether it makes systems open or closed.

□ □ Opportunities for Challengers

While standards are causing headaches for IBM, they offer opportunities for IBM challengers. Tandem president Jim Treybig, for example, believes that systems compatibility at Tandem has maximized savings both for customers and for Tandem itself. Treybig notes that all of Tandem's NonStop computers, from the very first one, have been designed to be fully compatible with one another. "We have an advantage because if we develop a new compiler, it runs on our inexpensive system, or the most expensive. IBM has to develop and support that for every system they have," says Treybig. "These basic similarities enable the same person to service $50,000 systems or $3 million systems."

The same economic advantage extends to R&D. If Tandem develops a new processor, it can run in the framework of all its systems. "So this architecture is absolutely unique," says Treybig. "The architecture has an economic advantage that's huge. It has inherent advantages for our customers because when they develop software to run on Tandem, it will run on any size system. If they want to build a network or make a change, they don't have to change anything, because all of our software is independent of the physical. This is a fundamental, inherent economic advantage for Tandem that I believe offsets any size advantage that IBM has."

Treybig continues: "We can pick areas and dominate them for our growth for a long time. IBM has to look at the whole world. So I don't think we are at a disadvantage due to size. We, in effect, are getting probably eight times more benefit in R&D due to this leverage. When we develop a new application generator we can compete with both the biggest IBM mainframe and the 4300, for one investment. They can't do that. They have to invest both places."

Tandem and other challengers are now focusing on moving to the third level of compatibility—compatibility with systems from other vendors. "Those companies who have successfully competed against IBM are working at the third level," says NAS's Martin. "Tandem is working very heavily up at this third level—it's very active in Corporation for Open Systems, very interested in this total interchangeability of systems. They can do it up there because they're not encumbered by base. Exploiting IBM's inability to move up to the level the customer really wants is a clear strategy for competing with IBM. I think you'll see many examples of that come out the next couple of years. You'll start seeing companies positioning up to third level."

Martin is making third-level compatibility a central part of NAS's strategy. "We are basically saying that whatever IBM does we'll do down at the first level and we'll stay compatible with an IBM-defined box. We're saying we'll do that, but we're also going to jump way up to the third level, and we'll interface with DEC, include Tandem stuff, and add Apollo workstations. We're doing this within the market segment of engineering/scientific today, so we're not trying to be all things to all people. But we tell customers we're going to make everything that they want to work at that highest level transparent in an engineering/scientific applied fashion."

Sun, more than any other company, owes its success to its enthusiastic embracing of open standards and compatibility. Not only has Sun based its machines on industry standards, but it has become the industry's leading cheerleader for open standards. The company has tried to convince its competitors of the value of adhering to industry standards in both hardware and software. Sun's vision is that computers should be sold like stereo systems, with all components conforming to widely accepted standards. As Sun president Scott McNealy told one reporter: "We are taking the idea of standards to a wild and crazy extreme."

On the software side, Sun has supported the leading graphics standards and communications standards. But perhaps most important, it has embraced AT&T's Unix operating system, and it is pushing to make Unix even more of a standard than it already is. Unix has been declared the operating system of choice on General Motors' factory floor, and the U.S. government requires Unix in vendor proposals. Unix has been adopted across the computing spectrum, from supercomputers to micros.

The acceptance of Unix as a standard has been driven by user demand for compatibility, not by IBM's stamp of approval. But Unix is gaining acceptance so rapidly that even IBM is now responding, to avoid losing its major customers. In late 1986, IBM made C, the Unix programming language, available on its System/370 processors, providing its huge user base with Unix portability. In a dramatic turnaround from the past, a standard was set and IBM was forced to follow in order to protect its customer base.

On the hardware side, Sun has used off-the-shelf standard technology (such as microprocessors from Motorola's 68000 series) from the start. Then, in 1987, Sun moved to establish a new hardware standard—a standard it wants to keep open, not proprietary. The standard is based on a Sun-developed microprocessor called SPARC (for scalable processor architecture). The SPARC chip uses the popular new design strategy known as RISC (for reduced instruction set computer). RISC technology makes the SPARC chip simpler and faster than previous microprocessors.

Sun has formed a powerful strategic relationship with AT&T. For its part, Sun agreed to work with an AT&T design team to refine Unix. In return, AT&T agreed to use the SPARC chip in its new computers. Now Sun is trying to enlist other companies to back the Unix-SPARC standard. In effect, it is *inviting* other companies to clone its work.

At first, industry analysts were skeptical of Sun's standards-oriented approach. "They said, 'You know, you're not providing anything new,' "recalls Sun's McNealy. "I said, 'Have you ever tried to glue Ethernet diskless node operation and bit-mapped graphics and Unix all together on a standard microprocessor and you can't change any of them from the user's perspective? Have you ever tried to glue all of that together?' And they said, 'No.' The analysts just didn't understand that it just took brilliant engineers to be able to put all that together and make it go fast. Our real proprietary advantage was the glue it took to piece all this together." Much to the analysts' surprise, users immediately saw the value in Sun's approach. After only five years of operation, the company was shipping systems at an annual rate of over $1 billion.

One key advantage to the Sun approach is that it eliminates the risk of lost investment for its customers. "Why would anybody want to buy from a little tiny startup computer company?" asks McNealy. "Price/performance was one reason. But they had to be safe; they had to have a fallback position. They had to have insurance if we flamed out." If Sun had taken the typical IBM approach and locked customers into a whole bunch of proprietary networks, CPU architectures, operating systems, and applications interfaces, then the customers would have been at great risk. If Sun had failed, customers would have lost all of their investment.

At its core, the standards movement is a security movement. Customers want and need security. By using industry-standard parts, Sun became a safe buy. Explains McNealy: "If we flame out, every other computer vendor in the business supports Ethernet; every other vendor in the business supports the 68000; every other vendor that's interesting to you supports Unix in an operating environment; every other vendor has the same windowing environment; every other . . . right on down the line."

❑❑ Succeeding in Open Environments

Today, almost everyone embraces the idea of open systems. Users want open systems because they want to mix machines from different vendors. And since users want open systems, vendors want open systems too. Open systems are becoming like Mom and apple pie— who could be against them?

But the story of open systems is not so simple. Open systems must be based on standards. But how does a standard become a standard? How can normally competitive vendors agree on standards? Which standards should vendors embrace? If some vendors try to gain control of a standard for their own benefit, what should other vendors do?

These are not hypothetical questions. Computer vendors are asking these questions every day. Already, there has been a backlash against Sun's promotion of a Unix standard. The problem is that there are many different versions of Unix. Other vendors agree that Unix should be a standard, but they don't want to let Sun and AT&T develop the Unix standard. They worry that Sun and AT&T might get some inside advantage. Sun, for all its altruistic talk about standards and open systems, might be trying to gain control of the new Unix standard.

So some companies are suggesting that some other version of Unix be used as a standard. A team of industry heavyweights, including Apollo, Digital, Hewlett-Packard, IBM, and Siemens, joined together in mid-1988 to form OSF (Open Software Foundation), chartered to offer an alternative. They want to build a more "open" Unix standard, based on an offshoot of Unix developed by IBM. Each company is chipping in several million dollars to help develop the competing Unix standard.

So life in the era of open systems can be pretty confusing. To help companies succeed in this new era, I can suggest only two general guidelines:

■ **Embrace open environments.**
■ **Keep something proprietary.**

On the first point, vendors have little choice. Users want open environments, so the vendors had better comply. In fact, it is a good idea to support new standards early. That way, you can help control the development of the standards. Moreover, you can take credit for driving the standard. Supporting standards is a way to demonstrate that you're on the side of the users.

On the other hand, companies can not compete on the basis of standards alone. Companies that live by standards can die by standards. Other companies, adhering to the same standards, could win on the basis of superior manufacturing technology. If companies do nothing but adhere to standards, then all computers will become commodities, and nobody will be able to make any money.

Thus, companies must keep something proprietary, something to differentiate their products. At first, this might seem paradoxical: proprietary architectures in an open environment. But there is no paradox. A company can adhere to industry standards on networking and basic operating systems, yet still offer proprietary features to distinguish its products. A vendor can distinguish its products in many ways. For example, it can offer an innovative user interface or higher reliability. That way, the vendor can offer the best of both worlds: special proprietary features, plus connectivity and basic compatibility with other brands of computers.

Indeed, some of the companies enjoying the most success today use proprietary architectures. Apple is the prime example. Inside Apple, there has been a long-term debate about retaining Apple's proprietary architectures. Had Apple bowed to the prevailing wisdom and converted to an MS-DOS operating system a few years ago, the company would have gone the way of the clones. But Apple chose a different path. The company maintained its proprietary operating system and user interface, but it enhanced its computers so that they can also run Unix and connect to the IBM world. In that way, Apple kept its machines unique, but it made them more valuable by opening them to a broader computing environment.

Jacques Stern, chairman of Honeywell-Bull, was on target when he said: "When you're competing with IBM, you must have the courage to be different." You must be different because uniqueness is what people pay for. You can't possibly offer any value-added if you're not different. If you're the same, the only thing separating your company from IBM becomes price. In order to compete, you'd always be selling at a price below IBM. When you're in that situation, you lose. In order to succeed, your value-added must be based on technology and customer-oriented solutions, not high-volume commodity.

Thus, standards will lead to diversity and creativity, not uniformity. Companies need to be more creative to distinguish their products in the new open environment. To succeed, companies must enhance their products regularly. They must continue to add new features that distinguish their products, while also staying within the framework of industry standards.

Practically the next time you look up, it's midnight, but you've done what you set out to do. You leave the basement thinking: "This is life. Accomplishment. Challenges. I'm in control of a crucial part of this big machine." You look back from your car at the blank, brick, monolithic Building 14A/B and say to yourself, "What a great place to work."

Tracy Kidder, *The Soul of a New Machine*

Putting It All Together

■ ■ YOU CAN LEARN a lot about a company by talking with taxi drivers. One time, after flying to Houston to visit Compaq Computer, I jumped in a taxi and asked the driver to take me to Compaq's headquarters. The driver knew very little about computers, but he knew about Compaq. "I don't know what they got going down at Compaq," he said, "but I've never seen a sour face coming or going from there."

Indeed, Compaq has put a lot of smiles on the faces of employees, customers, and investors. Compaq has become the model of a successful challenger company. Founded in 1982, Compaq reached $1 billion in annual sales faster than any company in history. Compaq has emerged as one of the three leading suppliers of personal computers, trailing only IBM and Apple.

In this chapter, I use Compaq as a case study to reinforce the ideas I've been discussing in the previous chapters. In particular, I focus on Compaq's approach to *leadership*, *relationships*, *innovation*, and *standards*. Compaq has put these ideas to work better than any other challenger; its story offers valuable lessons for all challenger companies.

❑❑ Establishing a Leadership Position

At first glance, Compaq's basic strategy seems crazy. From its very start, Compaq has gone head-to-head with IBM, selling IBM-compatible personal computers—computers that run the same software as IBM PCs and thus compete directly against IBM. But at every step, Compaq has carefully differentiated itself and its products, carving out a leadership role in the PC-compatible segment of the personal computer industry.

Compaq was founded by three engineers from Texas Instruments: Rod Canion, Bill Murto, and Jim Harris. The three had previously worked together as a team within TI: in 1979, they were asked to develop an IBM-compatible Winchester disk for TI. Together, they researched the competition and gradually developed a strategy. Then, they assembled a design engineering team and a marketing team—in effect, building a company within a company.

After TI's successful entry into the Winchester disk market, the three engineers decided to start a company of their own. At first, they thought about developing a new Winchester disk drive for the IBM PC. But prospective investors rejected their initial business plan, arguing that too many other companies were already crowding into that market.

So Canion, Murto, and Harris began exploring other ideas. One day, lunching at a Houston House of Pies, they came up with the idea of developing a portable computer compatible with the IBM PC. The key in differentiating the computer was its portability: at the time, IBM offered nothing but a desktop model. In classic entrepreneurial fashion, the three engineers quickly sketched a rough prototype on a napkin. Starting from that sketch, Compaq Computer Corporation was born.

Canion, the leader of the group, contacted venture capitalist Ben Rosen, who quickly embraced the idea of an IBM-compatible portable computer. Rosen helped Compaq line up other funding from the venture capital community. "It was kind of like the Northwest in the early days. There was a lot of wide-open territory," recalls Canion, who became president of the new company. "IBM just had one product on the market then: a dual-floppy PC. We felt like IBM would be busy filling out its basic PC line before they moved over into portables. Our product was not a directly conflicting product—it was more a companion product, really."

Canion continues: "Around the same time, there were other companies that came out with both portables and desktops. I'm convinced it was an advantage that we didn't have the desktop at the time because it would have positioned us more as a direct competitor with IBM. With only the portable, the dealers looked at us more as a complementary company and product. It really helped us get started. And a good start with a new company is what it's all about."

In positioning its product, Compaq relied on more than portability. It aimed to establish itself as a high-quality supplier. "We pioneered portability, but that wasn't enough, " says Canion. "That got us started, but people were already beginning to wonder when IBM was going to do a portable. We anticipated IBM's portable product being much like its PC, and we differentiated our computer even beyond just being portable. We improved on the display. We improved on the ruggedness by putting in a Winchester disk. And we used the styling as another differentiating feature. It was a different approach to a desktop personal computer."

As part of its effort to establish a high-quality image, Compaq decided to ignore conventional wisdom about pricing. "The dealers and everybody else we talked to recommended that we sell 20 percent cheaper than IBM. That was the folk wisdom," says Canion. "But we realized this was one area where we had to go against all recommendations. We were convinced that if we were going to stay in business, we were not going to do that." Indeed, Compaq presents a striking contrast to manufacturers of IBM clones. While the clone-makers have battled for low price and lost money doing so, Compaq has continued to add new features—and has kept its prices at or above IBM levels.

Compaq also set out to give *compatibility* new meaning. In the mainframe world, *compatibility* meant *similarity*. IBM-compatible computers could run *similar* software, but not the exact same software. Compaq wanted its customers to be able to use IBM PC software right out of the box. "Compaq was the first to define compatibility as true, total compatibility: you can run the same software, the same diskette," says Canion. "The redefinition is more revolutionary than people realized because the word *compatibility* has been around so long. We changed people's concept of compatibility."

The Compaq strategy paid off quickly. Compaq gained immediate acceptance as *the* high-quality supplier of IBM-compatible computers. In fact, dealers and customers thought of Compaq's initial product as "the IBM portable" because it was so compatible, so reliable, and styled in such a classy way. It could have just as easily been an IBM product—it was that kind of quality.

Compaq's first portable was shipped in January 1983. By the end of the year, the company had annual sales of $111 million—a record for a first-year company. Annual sales rose to $329 million in 1984. The following year, Compaq logged $504 million in sales, putting it on the Fortune 500 list after only three years in operation. Apple had taken five years.

❑❑ Relationships with Dealers

As they began developing a long-term strategy, Canion and his colleagues recognized that there were two things in the industry beyond IBM's control: IBM couldn't control the industry standard, and it couldn't control the dealer channel. "Those two things were our stakes in the ground," says Canion. "We built our whole strategy around differentiating in those areas where IBM couldn't change the rules on us."

In particular, Compaq set out to develop a strong set of relationships with computer dealers. In the personal computer market, relations with dealers are critical. More sales are gained or lost at the counter than anywhere else. For many years, I have used the maxim: "Keep the seller sold." Personal computer manufacturers must win not only a share of the dealer's shelf space, but a share of the dealer's mind.

Early on, Compaq's founders did extensive research into distribution channel alternatives. They focused on Xerox as a case study of how *not* to distribute a personal computer. "Xerox was heralded as the first major company to enter the PC business," recalls Canion. "About two months before IBM came in and did it right, Xerox shipped the 820 Information Processor, a CPM-based machine heralded as the machine that would take over the world. But Xerox went to every channel: to direct sales force, full-service computer dealers, hole-in-the-wall computer dealers, mail-order distributors, you name it. Their philosophy was to get as many 820s out as they could. As a result, very quickly, the least common denominator took over. The people who added value couldn't afford to compete with the guy down the street who added no value, or with the guy across the country who could mail to his customer at a rock bottom price. Eventually, all the channels offering service and support went away."

Based on the Xerox experience, Compaq executives wanted to avoid the conflicts that come with multiple distribution channels. They decided to pick a single channel. Direct sales was not a viable option. For products under $10,000, a direct sales organization would be an excessive expense. So Compaq settled on dealers. Dealers, if they do their jobs right, can provide better delivery than a direct sales force, and at a lower price. Dealers also provide system integration—they offer the best components from different companies, something that direct sales forces rarely do. Because dealers are independent, they continue to offer a choice of the best products. Thus, customers view dealers as reputable sources of information about the best products on the market. Moreover, dealer overhead for providing service, support, and training is lower than any other alternative. Says Canion: "A dealer done right can be the best distribution mechanism."

Compaq added value into the dealer channel by providing a type of support and understanding that IBM couldn't provide. Dealers have always watched IBM with a skeptical eye, viewing Big Blue as part supplier, part competitor. When IBM introduced its first personal computer, it set up a chain of company-owned retail stores, competing with the independent franchised stores. After a few years, IBM backed down in its retail store effort and began selling most of its PCs through independent dealers. But then dealers found themselves competing head-to-head with IBM on a new front. Both IBM and the dealers were trying to sell PCs to Fortune 1000 customers. IBM tried to reach these customers through its direct sales force, which accounted for 25 percent of all PC's sold. But dealers were also interested in the Fortune 1000 business. More and more retail dealers began building their own direct sales organizations, and they viewed Fortune 1000 customers as a lucrative market.

Thus, there was an inherent conflict in IBM's distribution strategy: dealers competed with IBM at the same time they were buying from IBM. IBM gave no clear indication of its future distribution plans, leaving the retailers to wonder exactly what role they would play. IBM's balancing act within the distribution channel created opportunity for other companies—and Compaq was the first company to capitalize on it.

Initially, Compaq was criticized for relying solely on dealers. "Conventional wisdom dictated that dealers would be just too big a limitation," he says. Now, however, Compaq is on the approved vendor lists of more Fortune 500 companies than any vendor other than IBM. "And we did it all through dealers," says Canion. "We didn't do it by going direct."

Throughout its history, Compaq has maintained solid relations with dealers. Compaq has never shown any interest in alternative distribution channels. In fact, in public forums, Compaq people are almost evangelistic about the retailers. Compaq executives have really listened to dealers and responded to their needs. For example, three months before Compaq was to release four new Deskpro computers, dealers complained that their stockrooms weren't big enough for four new models. Compaq responded. It redesigned the line so that the same chassis was used in all four models. Dealers then custom-designed the Deskpro by adding extra disk drives or hard disks, creating models ranging in price from $2,500 to $7,200.

Today, the difference in dealer relations between Compaq and IBM is not as great as it used to be, possibly because IBM has taken a few lessons in dealer relations from Compaq. But from Compaq's point of view, IBM can never erase the inherent conflicts between dealers and its direct sales force, nor can IBM fundamentally change its corporate attitudes. When IBM executives discuss products, programs, and pricing, they have to consider all distribution channels. "They have a very difficult negotiation job to make everybody happy," says Canion. "We have one channel to make happy. We think 'dealers.' Dealers are an integral part of everything we do: products, overall strategy, programs on how to promote the product."

❏❏ Sustaining Innovation

Many companies come to the market with a strong first product, grow quickly for a year or two, and then begin to falter. They never follow up on their initial innovation. Compaq never fell into this trap: it has sustained its growth through continued innovation. Throughout its history, Compaq has consistently been a step ahead of IBM, adding the incremental innovations that IBM has neglected.

Compaq's flurry of new products beginning in late 1986 provides a good example. First came the Deskpro 386. The 386 was the first personal computer to make use of Intel's powerful 80386 microprocessor. Among other things, the new machine allowed users to run all of the programs written for IBM PCs much faster, plus it provided desktop access to the software that had previously run only on minicomputers and mainframes.

With the Deskpro 386, Compaq broke the traditional pattern for clone-makers: rather than following IBM, it led IBM. Compaq brought the Deskpro 386 to market months before IBM introduced its first computer with the new Intel chip. At first, some people viewed Compaq's move as risky. After all, no one knew how IBM would use the new Intel chip. But after seeing all of the features on the Deskpro 386, customers decided not to wait for IBM. Sales of the new computer soared.

The success of the 386 played an important role in solidifying Compaq's position as an innovator. "As it became successful, as IBM didn't come out for a long time—in fact, nobody else came out for a long time—people recognized that Compaq was ahead on something very significant," says Canion.

Shortly after introducing the Deskpro 386, Compaq rolled out a new portable computer called the Portable III. "It surprised people because it wasn't expected," says Canion. "It was so much better than any other portable on the market, just head and shoulders technologically above anything else anybody had out. We really solidified our lead in the portable arena."

Compaq completed its new-product barrage three weeks later, when it introduced the Deskpro 286. The 286 was an advanced midrange computer, a notch below the 386 in price and performance. The new computer was not only ahead of where IBM was at the time, but it was ahead of new IBM products that came out later.

So in a six-month period, Compaq brought out innovative, leadership computers in three important product areas. Because IBM was expected to announce its own flurry of new products in April 1987, many industry observers predicted that the Compaq machines would get off to a slow start. They expected that customers would want to see the new IBM machines before making purchase decisions. But that didn't happen. March, the month before the IBM introduction, was Compaq's strongest sales month ever. "We really have entered a new phase," says Canion. "The 386 perhaps is the cornerstone of that transition, but all of the other pieces are important too. Compaq has emerged as a much stronger, much more solidly entrenched company, as the Number Two player."

Compaq has earned its reputation as a technology leader not through any single innovation, but through a long list of innovations. It has proved itself as a *sustained* innovator. Even before Deskpro 386, Compaq had a long list of accomplishments. Compaq offered the first fully compatible, fully portable PC. It offered the first monitor that could display both graphics and high-resolution text; with other computers, you needed two separate monitors. Compaq was also the first to integrate a reliable hard disk drive in a portable, the first to sell a fully compatible computer based on the Intel 8086 chip, and the first to offer a system with internal tape backup. "The list goes on and on," says Compaq vice-president Mike Swavely. "We have been a very constant source of innovation all along. It is just that now we are getting some of the recognition for that."

Swavely attributes Compaq's success in innovation, at least in part, to the company's team approach to development. There are no individual stars at Compaq. "It's discouraged to be a star in the sense of having a super ego, wearing your ego out on your sleeve, not respecting other people's contributions," says Swavely. "You can be a star at Compaq by delivering performance, but you won't necessarily be a star as would appear in other companies, because we just don't make room for big egos. We find that they very much get in the way of making good business decisions, so we try very hard to get that whole contentious environment out of the decision-making process."

The Compaq approach is almost diametrically opposed to IBM's approach. IBM is famous for setting up competitive situations among development groups within the company. Managers will assign two or more groups to work on similar projects, with an implied (or sometimes explicit) challenge: "May the best group win."

Swavely has heard a lot about the IBM approach, since three of the five people reporting to him used to work for Big Blue. "If you walk into a meeting at IBM and you are presenting a proposal, you win or lose, depending on whether your proposal is accepted or not. It's absolutely contention-driven," says Swavely. "The way they do business is to have multiple everything. You are always in a contention environment."

At Compaq, managers prefer to focus the competitive spirit on forces outside the company, rather than inside. "There is a great deal of competition, but not within Compaq," says Canion. "It's competition with our competitors. The enemy is not within Compaq. And that's critical. Once you start thinking of other groups within Compaq as the enemy, it's hard to change it."

Canion continues: "We're one big team and the enemy is IBM, the Japanese, whoever you want to pick. It's the competition. We drive hard to beat them. We see what they do, we anticipate where they might be at a certain time and with what technologies, and we want to beat them. Our people want to beat them. There's a great deal of a competitive spirit here. You want to direct your negative feelings outside the company, not inside."

❏❏ Relying on Standards

Compaq has capitalized on the changing nature of standards in the computer industry. Canion notes that today's standards, at least in the personal-computer market, are *industry* standards, not IBM standards. And that makes a big difference. Companies like Compaq do not have to sit back, worrying that IBM is going to change the standards overnight. That has freed Compaq to innovate, rather than simply following IBM's lead.

"We were not held back by IBM," says Canion. "We were able to leverage the industry standard and add to it without regard to what IBM was doing. The ability to move on to the next thing, and to open up new horizons without any penalty or any pain of separation from the past, is absolutely the most powerful force in the market today."

The biggest winners are the customers. In the old days, notes Canion, IBM was in total control of the rate of innovation. "If you look at IBM controlling the mainframe over a couple of decades, they just made improvements when they were ready," says Canion.

"It was to their benefit when they made changes, not necessarily when the users needed it."

The situation is very different today. Because of industry standards, customers can buy from non-IBM vendors without fear, uncertainty, and doubt. "Choice is probably the simplest way of describing it," says Canion. "Users are not dependent on one vendor to get around to doing things. It's a competitive environment so you've got people constantly pushing to do better and to improve and bring technology in."

When IBM does introduce new products and new "standards," customers no longer accept them uncritically. IBM doesn't automatically set the standards these days. Compaq and other IBM competitors now play an important role in helping to set new industry standards. Consider the case of the Micro Channel architecture in IBM's new PS/2 computers. With the Micro Channel, IBM tried to set a new standard, but Compaq rejected the new IBM standard. Instead, Compaq joined a group of IBM challengers in developing an alternative architecture that is more compatible with the familiar IBM PC standard. "IBM is taking away the benefits of that standard and giving customers a risky path to follow if they jump too soon," says Canion.

No one knows for sure which architecture will emerge as the industry standard for the future. But the industry will never be the same. IBM no longer calls the shots when it comes to setting industry standards. Says Canion: "It is clear to us that the industry is going to move forward with or without IBM, and we are going to be a major part of that."

❑❑ The Future

Despite Compaq's phenomenal success, there are still some doubts about the company's ability to survive in the long term. Since Compaq went public in December 1983, its stock has sunk as low as one-third its initial value. Some observers worry that Compaq, by selling through dealers, will have a difficult time staying in touch with the needs of its end users. Others worry that Compaq has no

TEN WAYS TO BEST IBM

1. Invest in a leapfrog idea or technology. Bet your company on advanced, market-creating technology. That sounds easy, but it takes courage, energy, and total commitment of the organization to drive the product through the organizational and market barriers.

2. Keep your development teams small, dedicated, and flexible.

3. Reward the technical people by listening to them and allowing them to become an integrated part of the marketing process.

4. Invest in something proprietary to differentiate your products, but also advance the cause of open systems.

5. Adopt industry standards early.

proprietary "hooks" in its computers. It offers no proprietary software or user interface. Compaq differentiates its products primarily on the basis of performance. So to maintain its leadership position, Compaq must continue to produce the highest performance machines in the industry—otherwise, customers will take their business elsewhere.

Finally, some analysts believe that the personal computer industry will eventually evolve into a commodity market, in which IBM-compatible computers will be differentiated by price alone. If that happens, no one making IBM-compatible machines will be able to make much money in the market.

Canion rejects these visions of the future. He argues that PC customers will increasingly use their machines as strategic weapons.

6. Establish a beachhead by dominating a niche. Learn to succeed and let everyone know you're a leader.

7. Stay in touch with the computer infrastructure: customer advocates, third-party software suppliers, value-added channels, industry consultants and pundits, and media people.

8. Stay profitable. Financial strength says more about your products' success than spec sheets do..

9. Keep your people fiercely competitive by constantly watching and analyzing the moves of your competition.

10. Seek to create your own markets and ways of doing business rather than trying to play the market-share game.

The ways in which a company uses its PCs could well determine whether the company succeeds or fails. Thus, buying decisions will be too important to make on the basis of price. "It is never just a price decision," argues Canion. "There will always be aspects of a computer purchase that will be important for business and that will allow companies to be differentiated."

Canion believes the new 386-based personal computers mark the beginning of an entirely new era in computing. "The 386 will give rise to a major acceleration," he says. "It will open doors to new workstation applications that have never been possible before. It will be the first true general-purpose desktop workstation." And Swavely says that the technology will continue to advance rapidly. By 1995, he predicts, PCs will process 16 million instructions per second, four times as many as today's best 386-based machines, and they will cost as little as $3,000. "That kind of performance was only available at a mainframe level three years ago," he says. "Applications that required mainframe horsepower won't require mainframe horsepower anymore. They will be done on your desktop with the right software and other support."

As Compaq offers these types of applications, its image will continue to evolve. "We have made a lot of progress over the last three years," says Canion. "Certainly, there is still a lot of hesitation and concern about competing with IBM. But today, we are in an infinitely different category than we were three years ago. At that time, there was total disbelief that anybody could succeed on the path we were choosing. And then, after quarter after quarter of delivering and executing, we have a lot of believers in Compaq's ability to do it. Our belief is that by continuing on that path, if we keep executing, we will gradually win more and more of the investment community over. If we are successful in delivering and gaining momentum with the 386 product, I believe the world will see Compaq very much differently than it does today. It still won't be a risk-free Compaq, but it will be a lot stronger than it is."

❑ PART THREE

The Future

The great obstacle to progress is not ignorance, but the illusion of knowledge.

Daniel Boorstin, *The Discoverers*

□ 11
Looking Ahead

■■ TRYING TO PREDICT the future is a risky business. Volumes have been written about the future of the computer industry—and few of the predictions agree with one another. Undoubtedly, some of the predictions will be right, but the vast majority will be wrong. The only thing that can be said with any assurance is that the future will bring more change. The computer industry of the future will be significantly different from the past, and in unpredictable ways.

It is the ever-quickening pace of innovation and technological change that makes the future of the computer industry so difficult to predict. Historian Arthur Schlesinger once noted that human beings have been on earth for about eight hundred lifetimes. Most of that time, they lived in caves. Movable type came into being only eight lifetimes ago, and industrialization appeared only three lifetimes ago. There have been more scientific and technological achievements in the last two lifetimes than in the first 798 combined! And there will probably be an even greater number in the next lifetime.

As a Dutch businessman once said in response to a Conference Board survey, "The future isn't what it used to be." In our ever-changing, technological world, we can no longer look at the future as an extrapolation of the past. The computer industry is particularly resistant to forecasting. In the late 1970s, few people had any idea of just how big the personal computer business would be in the 1980s. The most optimistic forecasts in the late 1970s projected a $2 billion market for personal computers by 1985. In fact, the market grew to more than $30 billion. Now there are more than 240 suppliers of personal computers, more than 150 PC clones, more than 60 companies making disk drives, more than 400 word-processing packages, more than 350 spreadsheet and database packages.

Technology has a way of creating new products, new ways of doing business, new markets, and new segments of old markets that even the most experienced industry participants do not (in fact, cannot) foresee. I prefer to look at the future and say, "We have yet to see or imagine what is possible." In the next ten or twenty years, we will see an information-intensive world turned upside down by the use of new computing tools and the forces of competition. The definitions of *computing power* will continue to change as new waves of technology take hold. But more that that, new markets, new ways of applying computer power, and new ways of doing old things will continue to emerge and surprise us. Media guru Marshall McLuhan said that "any technology gradually creates a totally new human environment." And indeed, computers will fundamentally change our total human enviroment.

The pace of change, and the uncertainty of its outcomes, will almost certainly increase in the future. What does all this change and uncertainty mean to today's computer companies? It means that moving into the future is more challenging than ever before. It means that the industry will look entirely different tomorrow than it does today. It means that managements must change their ways of thinking about how to compete. They must develop adaptive organizations, experiment more, integrate customers more intimately into the design process. They must offer solutions, not just products. They must not sit back and study the market, but rather help drive the market, pushing innovation, creating new solutions, developing new market niches, always adapting to the market's dynamics.

Having warned of the dangers of making predictions, I will now spend a little time making some predictions. But I will not try to predict particular events or products or technologies. I will not predict which companies will succeed or fail. Rather, I will discuss general trends that are likely to result from the continuing onslaught of technological change. In particular, I examine the future from four different perspectives: the future of the industry as a whole, the future of Japanese competition, the future of IBM, and the future of the IBM challengers.

❏❏ The Future of the Industry

The computer industry is often compared to the auto industry in the early part of this century. At that time, there were dozens of small

COMPUTERS AND THE FUTURE

- Increasing specialization of computers will create customized everything.
- Products and services will become increasingly personalized.
- Increasing options and choices will change customer habits and competitive marketing.
- Computers will become more transparent to users, spreading applications into uncharted territories.
- Small, entrepreneurial companies will create new computer applications that present-day producers have not yet even imagined.
- Big, bureaucratic companies will have increasing difficulty dealing with the pace of technological change and the fragmentation of markets.
- New technologies will cause managements to experiment with new forms of organization, management, marketing, and alliances.
- Computer literacy and technology transfer will accelerate in the coming century, causing increased competition from newly industrialized nations.
- The revolutions in computer science and biology will become intertwined, as computers are used for molecular modeling and biologists begin producing "biochip" computers.
- Everything will become digital. This convergence of technologies (audio, video, and computation) will lead to a convergence of industries (computers, electronics, communications, publishing, and entertainment).
- Standards will proliferate, continually change, multiply, and become totally confusing to the user.

companies in the auto business. But eventually, economies of scale drove all but a few of the companies out of the business, leaving the three auto giants of today.

Many people predict that the computer industry will follow the same course. Indeed, most of the industry leaders I interviewed predicted that only a handful of computer companies will survive into the twenty-first century. And, of course, all believed that their own companies would be among the survivors.

Listen to some of their statements.

John Cunningham: *You are going to see a lot of consolidation. You probably are going to end up with seven or eight or nine major players in the business—major being $10 billion companies and above—going out ten or fifteen years.*

Edson de Castro: *I think we are going to be in for a period of consolidation. That doesn't mean there aren't going to be some entries. But by and large, if we look at the year 2000 as a checkpoint, I think we're going to have been through a period of substantive consolidation. There will probably be six or seven or eight computer companies of all types that are going to make it. There are always going to be small companies making a small number of something. If you said maybe $10 million or $20 million of revenues as a threshold, I don't think there are going to be more than six or eight companies above that at the turn of the century.*

These arguments might seem reasonable, but I think that they are way off target. The computer industry is unlikely to follow the auto scenario any time soon. As we move ahead into the twenty-first century, the number of computer companies will continue to proliferate. Why? There are several reasons:

∎ The cost of entry into the computer business is relatively low and continues to decrease. By contrast, the cost of entry into the auto business has increased throughout its history. You can't enter the auto business by first entering the business of highway construction or glass making or tire making or oil refining. But in the computer business, you can enter from anywhere. You can enter from the chip market, the peripheral market, the software market, or the network and communications market. In the computer business, unlike the auto business, the pieces are often more important than the whole.

■ The demand for specialized computer applications continues to grow. Thus, large computer vendors find it increasingly difficult to be all things to all people. The auto business, by contrast, was built on mass production and mass market. Production was standardized to provide low cost and high volume. There were clear economies of scale.

In the computer industry, fragmented markets allow small companies to gain a foothold in the business. Companies can use narrow niches to get a start, gradually developing the resources needed to compete in broader areas. IBM, with annual revenues of almost $50 billion, was unable to stop the march of significantly smaller companies like DEC, Apple, and Compaq. Until large companies find a way to serve diversified, small market niches economically and efficiently, small companies will continue to flourish and gain access to the growth markets of the future. Consolidations will occur, but new companies will replace the ones lost to consolidation.

■ The computer business is a problem-solving business, not a producer of "iron." Computer companies that compete on the basis of low-cost hardware are doomed to fail. Today's consumers want computers that solve their specific problems. Successful computer companies are increasingly becoming service businesses. They provide solutions to problems, integrate the various functions of an office or business, improve productivity, improve competitiveness, cut costs, and add value.

Companies view the purchase of the corporate computer system very differently than they view the purchase of company cars or company trucks. The service aspect of the auto business is an after-market business, not integrated into the auto business itself. It is largely a fix-it business. The concept of computer service is much broader. Computer service is equivalent to constructing the roads and showing people how to drive safely—as well as maintaining the car and the roads, enhancing the engine at least every two years, and upgrading the driving rules. The computer business is a hands-on business.

■ Computer industry boundaries are constantly changing: new categories continue to appear and old categories merge together. In the auto business, there are only a few forms of transportation (cars, trucks, vans), and the uses of the vehicles are pretty much fixed. By contrast, new types of computers, and new uses of computers, are popping up every day and everywhere. A few years ago, I was having lunch with a journalist, discussing this very subject, and I mentioned that we still hadn't seen computers at tables in restaurants. Just then the waiter arrived with a computer in his hand to take our order.

For all these reasons, the computer industry is unlikely to follow in the footsteps of the auto industry. At the turn of the century, there won't be a Big Three dominating the computer industry. More likely, the industry will be driven by the Little 1000, a new set of innovative startups that will introduce new technologies, create new markets, and meet the specialized needs of tomorrow's computer customers.

❑❑ The Future Role of Japan

The Japanese have been so successful in so many technology businesses in the past twenty years that it is hard to imagine that they will be anything but successful in the future of the computer industry. Indeed, the Japanese computer industry has many things going for it. The industry has a well-developed and well-educated infrastructure. What's more, the industry can rely on strong support from the powerful Japanese semiconductor industry and the Japanese government.

The role of Japan's semiconductor industry could be particularly important. Although Japanese semiconductor companies have not been very successful in the important microprocessor market, they hold a tight grip on the market for semiconductor memory chips. For instance, they control more than 90 percent of the market for one-megabit dynamic random-access memory (RAM) chips. These chips are critical for computer manufacturers. In early 1988, a shortage of dynamic RAMs caused problems for every major U.S. computer company (with the exception of IBM, which produces its own RAM chips). Small and medium-sized computer companies, unable to buy sufficient memory, delayed shipments of their products, reducing sales and earnings. Some companies lost market position because of the shortage. This incident showed the degree to which the suppliers of memory chips—almost all Japanese—can control the fate of the computer industry. Japan could use its strategic position in the semiconductor market to leverage its position in the computer industry.

More than anything else, this RAM shortage demonstrates what leaders of the U.S. semiconductor industry have been saying for more than a decade: Japanese control of the semiconductor supply could control the future of the computer industry. This is the single most vulnerable aspect of the American computer industry. U.S. companies have the capital, the markets, and the knowhow—but not the chips.

The Japanese government is also playing an important role in supporting the growth of the country's computer industry. The government is pumping millions of dollars into research projects aimed at building the Japanese computer industry into a world leader in the next century. The government-funded Fifth Generation Computer Project, for example, aims to leapfrog the Japanese industry over the United States in the pivotal field of artificial intelligence, setting the foundation for twenty-first century "machines who think." In all, Japan spends about $50 billion on research and development. That's only one-third as much as the United States. But Japan does not carry the burden of defense-related R&D. In nonmilitary R&D, the two countries are roughly on a par.

So Japan has a long list of strengths. But don't expect the Japanese to rocket to the top of the computer market as they have done in other high-technology markets. Indeed, the next twenty years could be as difficult for Japanese computer companies as for IBM. Here are a few reasons why:

■ Japanese computer companies, like IBM, tend to be large bureaucratic organizations. As I have discussed, size and bureaucracy can be major disadvantages in today's fragmented, ever-changing computer market.

■ The Japanese have no successful venture-capital industry to sponsor new and different technologies. Although some people argue that venture capitalists have hurt American technology by draining off talent from large companies, it has been venture capital that has financed almost every major technology advance in the computer industry since the integrated circuit. It was venture capital that funded Intel, Apple, Tandem, Compaq, Lotus, BusinessLand, and thousands of other innovative hardware and software businesses. If bringing new technology to the market quickly is one of the keys to competitiveness, the U.S. computer industry has an edge.

■ Japanese technology companies have generally relied on large, volume markets to gain economies of scale. Their first entry into a market is almost always based on price competition. That strategy was successful in the semiconductor market, where the Japanese gained a beachhead by churning out millions of memory chips. But the same strategy is unlikely to work in the computer market. The computer industry is a fragmented, service-oriented business, where a commodity mentality will certainly fail. Japanese companies must learn to deal with the diversity and fragmented structure of today's computer industry.

■ Japanese companies have a much stronger record in *process* innovation than in *product* innovation. Process innovation is particularly important in industries where manufacturing costs are the determinant of success. But in the computer business, where the product creates the market, product innovation is crucial. There are no stationary targets to shoot for; products and standards keep changing. It is not enough to follow someone else's standard and shave the price. Compaq didn't succeed simply by mimicking the IBM standards; in fact, Compaq has often introduced new technologies *before* IBM. Apple and Tandem have succeeded by setting their own standards. Successful companies must provide links and bridges to existing standards, but they must also distinguish themselves through product innovation.

■ Japan cannot succeed if it produces only computer hardware. It needs software companies to produce innovative programs, and third-party resellers to develop vertical market solutions. The computer business is becoming more like a service business. Companies must supply value-added solutions in applications, networks, strategic ideas, and software.

■ Technology transfer is no longer a one-way street. From the early 1950s through the mid-1970s, Japan acquired almost all of its

R&D from the United States and Europe, often by buying licenses to foreign technologies. Now Japan is transferring technology to other countries, sometimes unwittingly, as it opens plants throughout the world. In order to stay competitive, Japanese computer companies will find it necessary to locate more and more of their operations in the United States. I believe this will be the most important vehicle of technology transfer back to the United States.

■ Japan's protected domestic market, a major advantage in the past, is gradually opening up. Unlike many American technology companies of the past, U.S. computer companies are making an active effort to buy and sell in Japan. Sun, Apple, Convex, Cray, DEC—all the challengers are working hard at penetrating the Japanese market. Traditionally, Japan allows new technologies to gain access to Japanese domestic markets only until Japanese companies can develop and produce competitive products. Once that happens, foreign companies have trouble penetrating the Japanese market. U.S. companies want to compete on a level playing field— and American politicians are increasingly becoming aware that competitiveness in the computer industry is the key to creating new jobs and new industries, and maintaining our standard of living. I believe that American society is coming to the realization that government, business, and education are in the competitiveness battle together. We cannot succeed without everyone pulling the oars in the same direction.

■ Japan is now battling a host of economic and social problems. Economically, the cost of labor is rising, and foreign countries are stepping up their complaints about Japanese trade policy, forcing Japan to open its markets and share its newfound wealth. At the same time, social standards are starting to fray. Japanese youths, once willing to accept a life of conformity and discipline, are beginning to raise questions similar to those raised by young Americans in the 1960s. All of these trends are introducing new uncertainties for Japanese industry in general, and Japanese computer companies in particular.

Perhaps most important, Japan can no longer "sneak up" on its foreign competition, as it often did in the past. American computer companies are watching and learning from their Japanese competitors. This is very unlike the situation in the 1960s and 1970s, when U.S. companies in the consumer-electronics and semiconductor industries largely ignored the growing competition from Japan. Every American computer executive I talked to has a "Japanese policy." That is, they are actively monitoring the moves of Japanese competitors, and they spend a great deal of time planning how to compete with the Japanese. Almost all U.S. computer companies have borrowed ideas from the Japanese in the areas of manufacturing and quality control. Today, U.S. computer companies have manufacturing and quality programs on a par with the best in Japan.

American companies have also learned to think internationally, just like their Japanese competitors. U.S. computer companies are focused on international markets almost from day one: they see the world as their market. Tiny Convex, the minisupercomputer pioneer, ships almost 50 percent of its product to Europe and Asia. The company buys parts, develops and manufactures products, and provides support in all the major countries of the world.

To be sure, Japanese computer companies will continue to grow, but it will not be easy for them to take over markets. This time, American companies are ready.

□□ The Future for IBM

Many people I interviewed predicted that IBM will remain the dominant force in the computer industry. But in my mind, that is hardly a sure thing. In the past, many industrial giants have faded from leadership as technologies and markets shifted. The railroad industry, the steel industry, the textile industry—all of these industries have seen powerful companies crumble in the face of change and competitive innovations. As one computer industry analyst told the *New York Times:* "The companies that benefited the most from the old world have the most to lose in the new world."

IBM's top executives have begun to recognize that many of the company's strategies and structures are outdated and even counter-productive in the new computer industry. In the last year, IBM has made numerous adjustments in an effort to make the company more flexible and responsive. For example:

■ To move into the supercomputer market, IBM decided in late 1987 to go outside its own labs and fund an entrepreneurial startup group headed by maverick designer Steve Chen. "They'll give us technology and participation and then leave us alone in the Wisconsin woods," Chen told the *Los Angeles Times*. "Staying separate from IBM is a very significant point for me. They understand that if we were to merge we could lose our creativity and ability to move fast."

■ In January 1988, IBM announced a sweeping restructuring of its organization, including the creation of five new and autonomous units. This move aimed to unclog Big Blue's infamous bureaucracy and to decentralize decision making, allowing IBM managers to implement decisions more quickly and effectively. IBM chairman John Akers described the reorganization in historic terms: "This is a fundamental change in the way we do our business, as significant as any we have ever made. If it works, it will make our employees more entrepreneurial, more accountable, and more independent."

■ In June 1988, IBM introduced a new line of midrange computers to replace its aging System 36 and System 38 computers. The new computers, code-named Silverlake and officially called Application System/400, were designed to compete more effectively with Digital's VAX minicomputers. Previous IBM attempts to compete with the VAX fell short of expectations, partly because of lack of software. But this time, IBM worked closely with software developers to make sure that a full range of software was available by the time the new computers were introduced.

■ Long known for its secrecy, IBM is experimenting with its own version of *glasnost*. More than ever before, IBM officials are meeting with journalists, consultants, and customers, discussing company strategies and product directions.

■ IBM is aiming to bring some consistency to its collection of incompatible systems. In 1987, IBM announced that it is developing a new software environment, dubbed Software Application Architecture (SAA), that will allow *all* IBM computers, from personal computers to mainframes, to "talk" with one another and share software easily.

These changes, though dramatic, are unlikely to change IBM in any fundamental way. Evolutionary change can take a decade or more. IBM's predicament is perhaps best understood through the lens of futurist Alvin Toffler. Toffler divides the history of technological change into three great "waves." Each wave is characterized not only by new technologies, but by dramatic changes in social development and organizational structures. The first wave was marked by the rise of agriculture more than ten thousand years ago. Rather than living in migratory groups, surviving on herding and hunting, humans began to cultivate land and live together in villages and settlements. The second wave began about three hundred years ago. It brought industrialization and factories and new transportation forms and giant urban areas. Toffler's third wave, beginning in the last few decades and picking up momentum now, is marked by (among other things) the rise of computers, the new status of information, and the internationalization of markets and technologies.

Apple chairman John Sculley views IBM as a great second-wave company, but questions whether IBM can adjust to the third wave. "IBM is the epitome of everything that establishment corporate America respects," explains Sculley. "It is a superbly managed company. It has the customer at the top of the list of every priority that the company has. Its credibility is founded upon its stability and the assurance that IBM will always be there to meet its customers' needs. And those things are all very positive. No one can quarrel with that. It's really the role model for how to run a great second-wave corporation. But how does that work when suddenly your lens is no longer the industrial age where the United States is at the top of the hierarchical economy?"

Sculley argues that the world economy is changing in fundamental ways—ways that cut against IBM's strengths and IBM's culture. "We're now in this global dynamic network where the United States is just one of the players," says Sculley. "And where competition can't be defined by how large a company is, how much economy of scale it has, how much self-sufficiency it has, how much stability it has, or how much tradition it brings with it. It is really defined in terms of how quickly can it respond to the changes of this global dynamic network. How can it compete with competitors who may not even exist for another five years?"

Sculley believes (and I agree) that the challenges IBM faces are the same ones that all large institutions face today. He notes that small organizations generally have a clear vision, often set by the founders themselves. "Yet as a company gets larger and more institutionalized," says Sculley, "things are all broken up into little pieces and delegated out throughout the corporation so that it's almost impossible for any manager to be able to be in total control of defining what that overriding vision is going to be."

Sculley continues: "My sense is that probably the only way that institutions ever change is when they have to go through crisis. IBM has gone through some tough times in recent years, but they haven't gone through a real crisis. I think they have very smart people at IBM. They are all well aware, obviously, of what their expenses are. They're well aware of the kind of competition they have. But there are just a limited number of things they can do without radically changing the very foundations upon which that corporation is built. It's almost like a president of the United States can see what he wants to do, but his hands are tied unless someone goes back and redefines the Constitution. Now who is going to redefine the constitution in the case of these big institutions?"

What does that mean for the future of IBM? As I said in the introduction, it's hard to imagine a world of computers without IBM. But the computer giant is increasingly vulnerable, susceptible to intelligent attack. The computer industry—and the world it serves— will change dramatically between now and the end of the century, and those changes will not play to IBM's strengths. Increasingly, the computer industry will be driven by challenger companies that are trying to build a new future, rather than giants that are trying to protect their past.

❏❏ The Future for IBM Challengers

Most of us laugh when we read the predictions of Jeane Dixon at the beginning of each new year. But then we go back to our businesses and create five- and ten-year business plans. That type of planning just doesn't work anymore. No one can predict the future with any certainty. Looking into the future, we find only paradox. We can predict only unpredictability, and the only certainty is uncertainty.

At first, the unpredictability of the future might seem to imply that there is no way for companies to prepare for the future. After all, strategists thrive on predictions and certainties. But planning for an unpredictable future is not a hopeless task. There are many ways for companies to cope with uncertainty. Rather than shooting at particular targets, companies must learn to shift targets quickly and efficiently. Rather than adopting specific strategies, companies must learn to alter their strategies as technologies and markets change. Most important, companies must learn to take the future into their own hands. Rather than trying to fit into a fixed future, companies must help create a new future.

Cray chairman John Rollwagen tells a story about Seymour Cray, the brilliant computer designer and founder of Cray Research. In the early days of the company, Rollwagen used to take people—potential customers, financial people—to see Cray. People were so much in awe of Cray, at first they didn't know quite what to say. But inevitably they would ask Cray about the future. What changes did he expect? What trends did he see?

Cray would give a careful, detailed response. Rollwagen was always amazed. Cray's picture of the future was "consistent within itself all the way out to however far you wanted to go, consistent with where we were at the time, consistent with everything that had led up to then. It was all in a nice package," says Rollwagen. "And inevitably, the visitors and I would leave saying: 'Well, that's it. We don't have to think about that any more. I mean, we know what's going to happen.'"

Then, one day, Rollwagen had a realization. He explains: "I noticed that I could bring someone back literally one day later, or two days later, and they'd inevitably ask the same question. Seymour would inevitably do the same thing, inevitably be consistent all the way out with today, with the past, and be just this much different from yesterday. It's wonderful. He's always got a plan and he's always changing it."

The same might be said for all of the successful challengers: they all have plans, but they are always changing and adapting their plans. None of the successful challengers see the future as a stationary object that they are shooting for. It is a moving target—and they are helping make it move.

Certain people in the computer industry are called "visionaries." Bill Gates, Steve Jobs, Seymour Cray, Mitch Kapor, Bill Joy—we are often led to believe that these people see the future more clearly than the rest of us. But there is no such thing as a visionary. These people succeed not because they can see into the future with any certainty, but because they help *create* a new future. They have a vision of what *might* be—then they make it happen.

The future is not fixed: there are many possible versions of the future. For many challengers, it is their fear that the future just might go in another direction that drives them to work so intensely to make *their* future happen. They build their future by putting each piece in place, one piece at a time, always being willing to challenge the conventional wisdom. It isn't wild visionaries that IBM has to worry about. It's the people who put one foot in front of the other, day by day, not burdened by the past.

The successful challengers of the future will be companies that thrive on change. Rather than sitting back and letting it happen to them, these companies will become catalysts of change. They will influence change to guarantee themselves a piece of the future. They will build adaptability to change into the structures of their companies. In short, they will take the steps that are needed to survive and thrive in an ever-changing world of computers.

Sources

Introduction

page 7
Thomas Watson Jr., quoted in *Fortune*, August 31, 1987, pp. 24 ff.

Chapter 1   Growth of a Giant

page 14
For more on the history of IBM, see: Thomas Watson, Jr., *A Business and Its Beliefs* (New York: McGraw-Hill, 1963); and Richard DeLamarter, *Big Blue: IBM's Use and Abuse of Power* (New York: Dodd, Mead, 1986).

page 20
Thomas Watson Jr., *A Business and Its Beliefs* (New York: McGraw-Hill, 1963).

pages 22–23
Buck Rodgers, *The IBM Way* (New York: Harper & Row, 1986).

Chapter 2   The Game Changes

page 33
Apple 1984 television commercial. Reprinted by permission of Bruce Mowery, Apple Inc.

page 34
John Verity and Geoff Lewis, "Computers: The New Look," *Business Week*, November 30, 1987, p. 112.

Chapter 3   Big Blues for Big Blue

page 54
Andrew Pollack, "The Daunting Power of IBM," *New York Times*, January 20, 1985; "Personal Computers: And the Winner Is IBM," *Business Week*, October 3, 1983, pp. 76-79.

page 66
Daniel Boorstin, *The Discoverers* (New York: Random House, 1983); "History Teaches 'We Don't Know What We Think We Know,'" *U.S. News & World Report*, March 5, 1984, p. 73.

page 73
"Blue Battlers: Winning on IBM's Home Court," *Electronic Business*, January 1, 1986, p. 28.

Chapter 4   Leadership

pages 80
Tom Alexander, "Cray's Way of Staying Super-Duper," *Fortune*, March 18, 1985, p. 66.

Chapter 5   Innovation

page 96–97
Donald K. Clifford, Jr., and Richard E. Cavanagh, *The Winning Performance* (New York: Bantam Books, 1985).

Chapter 6   Image

page 118
William D. Marbach, *Newsweek*, November 14, 1983; *Washington Post*, November 2, 1983; "IBM's 'Junior' Will Stabilize a Chaotic Market," *Business Week*, November 14, 1983, p. 49.

page 119
"System Review: The IBM PCjr," *Byte*, August 1984, p. 254.

page 120
Janice Castro, "Kicking Junior Out of the Family," *Time*, April 1, 1985, p. 58.

pages 121–122
Michael W. Miller, "Mystery Machine: IBM Computer Buyers Are Bewildered by PCs with Secret Ingredients," *Wall Street Journal*, March 22, 1988.

Chapter 8   Relationships

page 152
Information on Japanese acquisitions of licensing agreements is from James C. Abegglen and Akio Etori, "Japanese Technology Today," advertorial in *Scientific American*, October 1980, p. J20.

Chapter 9   Standards

page 164
Thomas G. Donlan, "Fixing the Tower of Babel: Helping Computers to Communicate," *Barron*'s, vol. 66, August 11, 1986, p. 14.

pages 172
Richard DeLamarter, *Big Blue: IBM's Use and Abuse of Power* (New York: Dodd, Mead, 1986).

Chapter 11   Looking Ahead

page 198
Arthur Schlesinger, "The Challenge of Change," *New York Times Magazine*, vol. 135, July 27, 1986, p. 20.

page 199
Marshall McLuhan, *Understanding Media* (New York: New American Library, 1964).

page 207
Andrew Pollack, "A New 'Fast Lane' in Computers," *New York Times,* February 23, 1988, p. D1,

page 208
Carla Lazzareschi, *Los Angeles Times*, December 27, 1987, part 4, p. 3.

page 209
Alvin Toffler, *The Third Wave* (New York: William Morrow, 1980).

Index